Distributed Control and Optimization Technologies in Smart Grid Systems

MICROGRIDS AND ACTIVE POWER DISTRIBUTION NETWORKS

SERIES EDITORS

Ali Davoudi
University of Texas at Arlington, USA

Josep M. Guerrero
Aalborg University, Denmark

Frank Lewis
University of Texas at Arlington, USA

PUBLISHED TITLES

Distributed Control and Optimization Technologies in Smart Grid Systems

Fanghong Guo • Changyun Wen
Yong-Duan Song

CRC Press
Taylor & Francis Group
Boca Raton London New York

CRC Press is an imprint of the
Taylor & Francis Group, an **informa** business

CRC Press
Taylor & Francis Group
6000 Broken Sound Parkway NW, Suite 300
Boca Raton, FL 33487-2742

First issued in paperback 2022

© 2018 by Taylor & Francis Group, LLC
CRC Press is an imprint of Taylor & Francis Group, an Informa business

No claim to original U.S. Government works

ISBN-13: 978-1-138-08859-7 (hbk)
ISBN-13: 978-1-03-233933-7 (pbk)
DOI: 10.1201/9781315109732

Visit the Taylor & Francis Web site at
http://www.taylorandfrancis.com

and the CRC Press Web site at
http://www.crcpress.com

Contents

SECTION III:DISTRIBUTED TERTIARY OPTIMIZATION   73

List of Figures

List of Tables

Preface

As the main building block of the smart grid systems, microgrid (MG) integrates a number of local distributed generation units, energy storage systems, and local loads to form a small-scale, low- and medium-voltage level power system. In general, an MG can operate in two modes, i.e., the grid-connected and islanded mode. Recently, in order to standardize its operation and functionality, hierarchical control for islanded MG systems has been proposed. It divides the control structure into three layers, namely, primary, secondary, and tertiary control. The primary control is based on each local distributed generation (DG) controller and is realized in a decentralized way. In the secondary layer, the frequency and voltage restoration control as well as the power quality enhancement is usually carried out. In the tertiary control, economic dispatch and power flow optimization issues are considered. However, conventionally both the secondary and tertiary control are realized in a centralized way. There are certain drawbacks to such centralized control, such as high computation and communication cost, poor fault tolerance ability, lack of plug-and-play properties, and so on. In order to overcome the above drawbacks, distributed control is proposed in the secondary control and tertiary optimization in this book.

In the secondary control, restorations for both voltage and frequency in the droop-controlled inverter-based islanded MG are addressed. A distributed finite-time control approach is used in the voltage restoration which enables the voltages at all the DGs to converge to the reference value in finite time, and thus, the voltage and frequency control design can be separated. Then, a consensus-based distributed frequency control is proposed for frequency restoration, subject to certain control input constraints. The proposed control strategy can restore both voltage and frequency to their respective reference values while having accurate real power sharing, under a sufficient local stability condition established.

Then the distributed control strategy is also employed in the secondary voltage unbalance compensation to replace the conventional centralized controller. The concept of contribution level (CL) for compensation is first proposed for each local DG to indicate its compensation ability. A two-layer secondary compensation architecture consisting of a communication layer and a compensation layer is designed for each local DG. A totally distributed strategy involving information sharing and exchange is proposed, which is based on finite-time average consensus and newly developed graph discovery algorithm.

In the tertiary layer, a distributed economic dispatch (ED) strategy based on projected gradient and finite-time average consensus algorithms is proposed. By decomposing the centralized optimization into optimizations at local agents, a scheme is proposed for each agent to iteratively estimate a solution for the optimization problem in a distributed manner with limited communication among neighbors. It is shown that the estimated solutions of all the agents reach consensus of the optimal solution asymptomatically. Besides, two distributed multi-cluster optimization methods are proposed for a large-scale multi-area power system. We first propose to divide all the generator agents into clusters (groups) and each cluster has a leader to communicate with the leaders of its neighboring clusters. Then two different schemes are proposed for each agent to iteratively estimate a solution of the optimization problem in a distributed manner. It is theoretically proved that the estimated solutions of all the agents reach consensus of the optimal solution asymptomatically. In addition, a novel hierarchical decentralized optimization architecture is proposed to solve the ED problem. Similar to distributed algorithms, each local generator only solves its own problem based on its own cost function and generation constraint. An extra coordinator agent is employed to coordinate all the local generator agents. Besides, it also takes the responsibility for handling the global demand supply constraint. In this way, different from existing distributed algorithms, the global demand supply constraint and local generation constraints are handled separately, which would greatly reduce the computational complexity. It is theoretically shown that under proposed hierarchical decentralized optimization architecture, each local generator agent can obtain the optimal solution in a decentralized fashion.

A distributed optimal energy scheduling strategy is also proposed in the tertiary layer, which is based on a newly proposed pricing strategy named *PD pricing*. Conventional real-time pricing strategies only depend on the current total energy consumption. In contrast to this, our proposed pricing strategy also takes the incremental energy consumption into consideration, which aims to further fill the valley load and shave the peak load. An optimal energy scheduling problem is then formulated by minimizing the total social cost of the overall power system. Two different distributed optimization algorithms with different communication strategies are proposed to solve the problem.

Authors

Fanghong Guo received his B. Eng. degree in automation science from Southeast University, Nanjing, China, in July 2010, M. Eng. degree in automation science & electrical engineering from Beihang University, Beijing, China, in January 2013, and Ph.D. degree in sustainable earth from Energy Research Institute @NTU, Interdisciplinary Graduate School, Nanyang Technological University, Singapore, in November 2016. He is currently a scientist in Experimental Power Grid Center (EPGC), Agency for Science, Technology and Research, Singapore. His research interests include distributed cooperative control, distributed optimization on microgrid systems, and smart grid. He received the 2015 National Award for Outstanding Self-financed Chinese Students Study Abroad in 2015.

Changyun Wen received his B. Eng. degree from Xi'an Jiaotong University, Xi'an, China, in 1983 and his Ph.D. degree from the University of Newcastle, Newcastle, Australia, in 1990. From August 1989 to August 1991, he was a research associate and then postdoctoral fellow at University of Adelaide, Adelaide, Australia. Since August 1991, he has been with School of Electrical and Electronic Engineering, Nanyang Technological University, Singapore, where he is currently a full professor. His main research activities are in the areas of control systems and applications, autonomous robotic systems, intelligent power management systems, smart grids, cyber-physical systems, complex systems and networks, model-based online learning and system identification, signal and image processing.

Dr. Wen is an associate editor of a number of journals including *Automatica*, *IEEE Transactions on Industrial Electronics* and *IEEE Control Systems Magazine*. He is the executive editor-in-chief of *Journal of Control and Decision*. He served *IEEE Transactions on Automatic Control* as an associate editor from January 2000 to December 2002. He has been actively involved in organizing international conferences playing the roles of general chair, general co-chair, technical program committee chair, program committee member, general advisor, publicity chair and so on. He received the IES Prestigious Engineering Achievement Award 2005 from the

Institution of Engineers, Singapore (IES) in 2005. He received the Best Paper Award of IEEE Transactions on Industrial Electronics in 2017.

He is a fellow of IEEE, was a member of IEEE Fellow Committee from January 2011 to December 2013 and a Distinguished Lecturer of IEEE Control Systems Society from February 2010 to February 2013.

Yong-Duan Song received his Ph.D. degree in electrical and computer engineering from Tennessee Technological University, Cookeville, USA, in 1992. He held a tenured full professor position with North Carolina A&T State University, Greensboro, from 1993 to 2008 and a Langley Distinguished Professor position with the National Institute of Aerospace, Hampton, VA, from 2005 to 2008. He is now Dean of the School of Automation, Chongqing University, and the founding director of the Institute of Smart Engineering, Chongqing University. He was one of the six Langley Distinguished Professors with the National Institute of Aerospace (NIA), founding director of Cooperative Systems at NIA. He has served as an associate editor/guest editor for several prestigious scientific journals.

His research interests include intelligent systems, guidance navigation and control, bio-inspired adaptive and cooperative systems, rail traffic control and safety, and smart grid.

List of Symbols

Algebraic Operators

A^T Transpose of matrix A
A^{-1} Inverse of matrix A
$\det(A)$ Determinant of matrix A
$P_X[\cdot]$ Projection onto set X

Sets

R Set of real numbers
C Set of complex numbers
Z Set of integers
N Set of nonnegative integers

Others

$\mathbf{0}$ Zero vector with a compatible dimension
$\mathbf{1}$ Vector with a compatible dimension and all elements of one
$\lambda(P)$ Eigenvalue of matrix P

Acronyms

AC	–	Alternating Current
AMI	–	Advanced Metering Infrastructure
CERTS	–	Consortium for Electric Reliability Technology Solutions
CF	–	Communication Fault
CL	–	Contribution Level
DG	–	Distributed Generator
DC	–	Direct Current
DCSCS	–	Distributed Cooperative Secondary Control Scheme
DPGM	–	Distributed Projected Gradient Method
DSM	–	Demand Side Management
ECC	–	Energy Consumption Controller
ED	–	Economic Dispatch
EMA	–	Energy Market Authority
ESS	–	Energy Storage Systems
EU	–	European Union
FACA	–	Finite-time Average Consensus Algorithm
FC	–	Fuel Cell
GA	–	Genetic Algorithm
HVAC	–	Heating Ventilation and Air Conditioning
IS	–	Isolation Switch
LB	–	Local Bus
MG	–	Microgrid
MGCC	–	Micro-Grid Central Controller
NE	–	Nash Equilibrium
OPF	–	Optimal Power Flow
PAR	–	Peak to Average Ratio
PCC	–	Point of Common Coupling
PD	–	Proportional-Derivative
PFC	–	Power Factor Correction
PI	–	Proportional-Integral
PID	–	Proportional-Integral-Derivative
PR	–	Proportional-Resonant
PV	–	Photo-voltaic
RTP	–	Real-Time Pricing
SDP	–	Semidefinite programming
SG	–	Synchronous Generator
SLB	–	Sensitive Load Bus
TG	–	Thermal Generator
UCR	–	Unbalance Compensation Reference
UPS	–	Uninterruptible Power Supply
US	–	United States
VUC	–	Voltage Unbalance Compensation
VUF	–	Voltage Unbalance Factor
WT	–	Wind Turbine

INTRODUCTION I

Chapter 1

Introduction

In this chapter, we first present the background, motivation and objectives of the research in this book. Then some detailed literature reviews on microgird as well as its state-of-art control and optimization strategies are conducted. Finally, the major contributions and organization of this book are summarized.

1.1 Background and Motivation

The demand for electrical energy is increasing rapidly. It is estimated that electricity demand will double between 2000 and 2030, with an annual growth rate of 2.4%, faster than the increase of any non-renewable energy source [1]. Hence more renewable energy sources are needed in future energy systems. The Renewables Portfolio Standards in the United States sets a goal of increasing the percentage of renewable energy sources to 33% by 2020 [2]. In Europe, the target is to raise the renewable energy penetration percentage from the current level of 20% to 50% by 2050 [3].

The renewable energy sources are usually located in a distributed manner, e.g., photovoltaic (PV) and wind, as opposed to the conventional large centralized power plants. This leads to a larger and more complex networked energy system. It is a large nonlinear highly structured system consisting of a number of interconnected distributed generators (DGs) or subsystems. It is difficult to employ a centralized controller to control such a large-scale system for many reasons such as limited communication capability among the subsystems as well as limited computation ability in one single controller. In order to handle this issue, a decentralized control method has been proposed to design a local controller for each subsystem. For simplicity and familiarity, each local controller is usually designed and implemented by ignoring the interactions from other subsystems and only using its locally available information.

This is essentially equivalent to imposing structural constraints on the centralized controller. Thus controllability is restricted by the decentralized approach and system control performances are deteriorated. One typical example of the consequences of the drawbacks of such control strategy is the widespread blackout of August 2003 in North America [4]. In that accident, each subsystem only focused on maintaining its own subsystem stability and transferred the extra load to other subsystems, which made the overload more severe and eventually caused a cascading corruption [5].

Note that communication techniques have advanced significantly and highly efficient communication networks are readily available in recent industrial infrastructures. It is natural ask, Will the system control performance be improved by letting the local decentralized controllers communicate with their neighboring controllers? The answer should be positive. In fact, such control strategy is referred to as distributed control, which has been widely studied and implemented successfully in many other fields such as process control [6], traffic control [7], and so on. In this book, we propose to apply distributed control strategies to energy systems, which allow information exchange among local controllers by establishing a communication network topology among them. In fact, distributed control strategies can be considered as a tradeoff between the centralized control and the decentralized control by combining their advantages. The comparison among centralized, decentralized, and distributed control scheme is illustrated in Fig. 1.1.

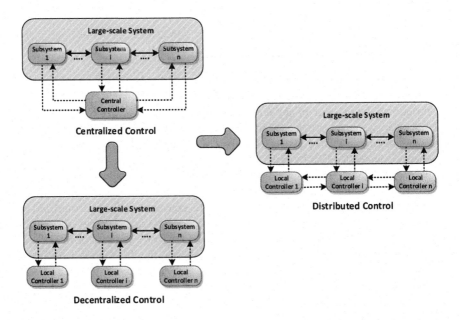

Figure 1.1: Centralized, decentralized, and distributed control scheme

There are also other factors that motivate distributed control [8], such as the deregulation of the electrical market. The electricity market is open for new suppliers and consumers can choose their own suppliers. The limitation that the current control methods imposes (low flexibility, low resilience to faults) also requires that the control algorithm for the power system should be reconsidered.

1.2 Objectives and Scope

Recently, the integration of the distributed renewable energy resources is evolving as an emerging and promising power scenario for the energy system. Smart grid, as a modernized electrical grid, uses information and communication technology to improve the efficiency, reliability, and economics of the production and distribution of electricity [9, 10, 11]. As the main building block of the smart grid, microgrid (MG) integrates a number of local DG units, energy storage systems, and local loads together to form a small-scale power system. In this book, we will propose some distributed control schemes for the MG control. Recently, in order to standardize its operation and functionality, hierarchical control for islanded MG systems has been proposed. It divides the control structure into three layers, namely, primary, secondary, and tertiary control. Conventionally both the secondary and tertiary control are realized in a centralized way. However, with the size of the MG growing, it brings some disadvantages such as poor dynamics and high computational cost. Instead of using a centralized controller, we decompose the centralized controller into decentralized ones and allow them to communicate with their neighboring controllers. In this way, we can make the control system more flexible and also reliable.

The research objectives and scopes are detailed as follows:

1. Based on the MG primary control model, develop distributed secondary control schemes for the islanded MG. This control structure allows each local controller to communicate with its neighboring local controllers. The main function of the secondary control is to restore the frequency and voltage values to their nominal values while keeping the real power sharing accuracy no matter how the loads vary.

2. Develop a distributed secondary control scheme for the voltage quality enhancement problem such as voltage unbalance compensation. The proposed control scheme should be flexible and make the controller have the "plug and play" property.

3. Develop a distributed tertiary control structure to solve the economic dispatch and optimal energy scheduling problems in the smart grid system. Besides, as the energy system has become a interconnected large-scale system, a more proper multi-cluster optimization method should be developed.

In the following, we will give some overview of recent research on MG as well as its control strategies.

1.3 Microgrid

An MG is a small-scale low and medium-voltage level power system that integrates a number of local distributed generator (DG) units, energy storage systems and local loads together [12, 13]. Different from the conventional synchronous generator (SG), the DG in the MG is inverter-interfaced with sustainable prime energy sources such as fuel-cells (FC), photovoltaic (PV), and wind power generators [14, 15]. Compared to the traditional fossil-fuel-based power grid, benefits brought by MG include less carbon consumption, faster demand response, and so on.

There are different types of MGs reported in the literature. According to the bus type, the MG can be classified into three types, namely, alternating current (AC) MG, direct current (DC) MG, and hybrid AC/DC MG [16]. More detailed information can be found in [16, 17, 18]. In this book, we mainly focus on the research of AC MG.

There are several reported MG-related research and development projects around the world. In Singapore, Energy Research Institute @ NTU (ERI@N), Nanyang Technological University, is setting up Southeast Asia's first and largest microgrid located at Semakau Island. It will demonstrate how to generate electricity from multiple sources including solar, wind, tidal, diesel, as well as integrate energy storage and power-to-gas technologies [19]. It is also reported that Energy Market Authority (EMA) in Singapore is piloting a microgrid test-bed at the jetty area of Pulau Ubin, an island north-east of Singapore [20]. This test-bed aims to assess the reliability of electricity supply within a microgrid infrastructure using intermittent renewable energy sources such as PV technology.

In the Europe Union (EU), the MG research has been conducted extensively since 1998. One recent project titled *More Microgrids: Advanced Architectures and Control Concepts for More Microgrids* aims to investigate alternative MG control strategies and alternative network designs, and develop new tools for MG management operation and standardization of technical and commercial protocols [21]. This is a follow-up project of the project *Microgrids: Large Scale Integration of Micro-Generation to Low Voltage Grids*, which is the first activity at EU level dealing in-depth with MG [22].

In the US, one well-known MG project is conducted by Consortium for Electric Reliability Technology Solutions (CERTS). This project explores the implications for power system reliability of emerging technological and environmental influences. It is reported that this CERTS MG concept has been fully developed and a laboratory-scale test system has been built [23].

Compared to the traditional power system, MG has the following advantages [24]:

1. It integrates the distributed renewable energy resources, thus it leads to less carbon consumption and also reduces energy cost.

2. It is more energy efficient. Since the DGs are usually close to the loads, the power transmission loss will be greatly reduced. Also it has faster demand response than the traditional power system.

3. It can provide high power quality. As the DGs are usually interfaced with

inverters, it not only provides the regulated AC voltage, but also has the ability to compensate for unbalanced and harmonic voltage.

4. It has improved utilization of conventional energy sources because there is less real, reactive, unbalance, and distortion power flowing through the distribution line.

However, there also exist some challenging problems that need to be solved in order to achieve and ensure the above advantages [24]:

1. As the prime renewable energy sources are intermittent and uncertain, they have irregular power injection, so power flow regulation and peak power shaving strategies are needed. Also it requires MG equipped with energy storage devices.

2. The MG distribution network is a weak and inertia-less grid. Each DG has non-negligible internal impedance; also the inverter has negligible physical inertia, which makes the system potentially susceptible to disturbances, so it has poor frequency and voltage stability.

3. As the power flow in the MG is bidirectional, the conventional voltage stabilization techniques are not applicable, new control and protection strategies are needed.

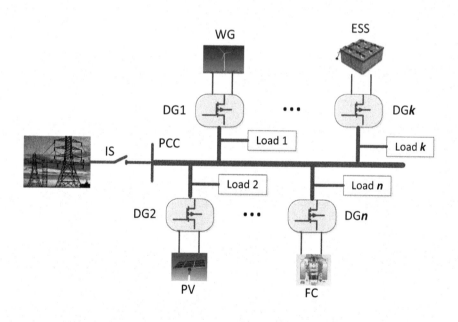

Figure 1.2: General structure of the MG

Generally, the MG can operate in two modes, i.e., the grid-connected and the islanded mode [25]. In the grid-connected mode, the MG is connected to the main power grid by closing the isolation switch (IS) in the point of common coupling (PCC), shown in Fig. 1.2. Due to larger capacity of the main grid, the frequency of the MG is dominated by the main grid. Also, the power mismatch of the MG system can be immediately covered by the main grid. In this mode, the MG system dynamics is fixed to a large extent by the main grid because of the smaller size of the DG units [26]. The DGs in this mode act like current sources, aiming to deliver scheduled constant real and reactive power to the grid. However, when a fault occurs in the main grid, the MG needs to open the IS to protect itself and then operates in the islanded mode. In this mode, the MG should be able to handle the following issues [27]:

1. Voltage and frequency management. The MG should maintain its frequency and voltage to certain nominal values with acceptable errors.

2. Supply and demand balancing. The MG should re-dispatch the real and reactive power among the DGs and loads while keeping the supply and demand balanced.

3. Power quality. It contains two levels. The first is reactive power and harmonic voltage compensation at each DG's output terminal, and the second level is reactive power compensation, unbalanced, and harmonic voltage compensation at the PCC.

1.4 Control Strategies of MGs

According to the aforementioned operating modes of MGs, the control strategies of the MGs can be classified into two different types.

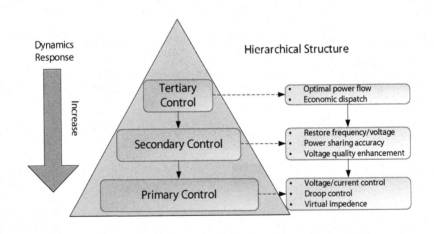

Figure 1.3: Hierarchical structure for the islanded MG control [28, 29, 30]

In the grid-connected mode, DG acts like a current source, and it is also called *PQ* control strategy [26]. The current components in phase (i_{act}) are responsible for the control of real power P; while the quadrature component of the current (i_{react}) is for the control of reactive power Q. This control strategy is quite straightforward. The real and reactive power output is regulated to the predetermined reference values.

While in the islanded mode, the control of the MGs becomes much more complicated and challenging. Recently, in order to standardize their operation and functionalities, hierarchical control for islanded MGs has been proposed [28, 29, 30]. As shown in Fig. 1.3, it divides the control structure into three layers, namely, primary, secondary, and tertiary control. The primary control is based on each local DG controller, which mainly consists of voltage loop, current loop, and droop control function [31]. As there is no information exchange between them and they are totally decentralized, the DGs share the real and reactive power autonomously by using the well-known droop method. The droop method emulates the droop characteristics of the traditional synchronous generators and regulates the frequency and voltage output based on local real and reactive power generation. However, there are some drawbacks in the primary control layer [32, 33]. Firstly, as the droop control function may cause the frequency and voltage deviation, it cannot guarantee zero frequency and/or voltage regulation errors. Secondly, the real and reactive power sharing accuracy is deteriorated when the ratio of line resistance to line reactance is high. Then a secondary control is proposed and applied to solve such problems. Its main function is to restore the frequency and voltage to their nominal values. Additionally in [34, 35], appropriate methods are also proposed to enhance the voltage quality including compensating for the voltage unbalance and harmonic distortion in the secondary control. In the tertiary control, economic dispatch, power flow optimization, and optimal energy scheduling issues are usually considered.

In this book, we mainly focus on the MG control in islanded mode. Hence, in the following, we will give a detailed literature review of this hierarchical control strategy.

1.4.1 Primary control

In this layer, the primary control is implemented on each local DG. Its main function is to regulate the frequency and voltage output of the inverter. In most existing literature [26]-[36], the inverter operates analogous to an uninterruptible power supply (UPS) and it has inner current control loop and outer voltage control loop.

Besides, the power sharing controller (droop controller) is always applied in the primary control to determine the reference voltage $v_o^\star$. It mimics the property of the conventional SG and behaves as "the virtual inertia." The frequency (voltage amplitude) decreases when the real (reactive) power increases respectively. In order to improve the performance of the conventional droop controller, several other modified droop controllers have been proposed in the literature [37, 38, 39]. More detailed introduction can be seen in the review paper [30].

It should be noted that such droop controller is based on the fact that the output and transmission impedance are inductive. However, the R/X ratio in the current MG, especially for the low-voltage or middle-voltage level MG, is not small enough. This leads to a poor decoupling between the real and reactive power control [40]. Recently, in order to solve this problem, the concept of "virtual impedance" has been proposed in [41, 42, 43]. The output current i_o "goes through" the virtual impedance and feedbacks to the reference voltage v_o^*. In this case, by properly choosing the value of the virtual impedance, we can make the inverter output impedance be our expected value. In contrast to physical impedance, this virtual impedance has no power losses.

The above introduction forms the main structure of the primary control. Roughly speaking, the primary controllers can be implemented in three frames, i.e., dq frame (the rotating frame), $\alpha\beta$ frame (the stationary frame), and the abc frame (the natural frame). How to choose a proper control frame is another research topic, and it will not be discussed in detail in this book. Besides, to improve the transient response of the primary control, proportional-integral-derivative (PID) [29], proportional resonant (PR) controllers [35] are utilized in both the current and voltage control loop. Generally speaking, PID controller is used in the dq frame while the PR controller is implemented in the $\alpha\beta$ frame.

In spite of the aforementioned analogous UPS control method, there also exists another inverter primary control method, which is based on the concept of *synchronverter* [44]. This concept is to make the inverter behave like an SG. Its model is derived from the conventional SG model, so that it can be easily embedded in an MG or a power system with many conventional SGs. Also it offers some advantages over the conventional control strategy as it introduces controlled frequency dynamics as well as the emulated inertia. Some conventional SG control strategies can be directly applied to this model. In [45], a nonlinear MG stabilizer for the synchronverter is designed via the adaptive backstepping technique.

Further, it is reported that there also exist many other advanced control methods in the primary control. These methods include the sliding mode control [46], Lyapunov function-based control [47], model predictive control [48], robust control [49, 50], and deadbeat control [51]. The main objective of all these methods is to regulate the voltage and current to track their desired values.

1.4.2 Secondary control

In this subsection, we will mainly review two control issues in the secondary control layer. The first one is a frequency and voltage restoration problem; the other is a voltage quality enhancement problem.

As discussed before, the implementation of the droop control function in the primary control may cause frequency and voltage deviations, especially when the heavy loads are connected to or disconnected from MGs. In order to compensate for these deviations, secondary control is introduced. The dynamics response in this control hierarchy is designed to be much slower than the primary control. In fact, this control layer is similar to the load frequency control (also called automatic generation control) in traditional power systems.

Conventionally, a centralized secondary controller called the MG central controller (MGCC) is designed in this layer [29]. The frequency of the MG and voltage of each DG are sampled and then compared with the corresponding reference values ω^{ref}, E^{ref}, respectively. Then the output control signals of the secondary control are generated by PI controllers, respectively. The control laws in [29] are listed below:

$$\delta\omega = K_{P_\omega}(\omega^{ref} - \omega) + K_{I_\omega}\int(\omega^{ref} - \omega)dt + \Delta\omega_s \qquad (1.1)$$

$$\delta E = K_{P_E}(E^{ref} - E) + K_{I_E}\int(E^{ref} - E)dt \qquad (1.2)$$

where K_{P_ω}, K_{I_ω}, K_{P_E}, and K_{I_E} are the PI gains. $\Delta\omega_s$ is an additional term to compensate for the frequency deviation between the MG and the main grid. If the MG is disconnected from the main grid, this additional term disappears, i.e., $\Delta\omega_s = 0$.

However, this method has certain intrinsic disadvantages such as poor control dynamics, high communication burden, and lack of robustness to communication failure. A distributed control structure can be applied in the secondary control, namely each local DG controller can communicate with its neighboring DGs. In this way, even if communication failure occurs in some DGs, it will not affect other DGs [52, 53]. This kind of distributed control strategy has been reported in some of the literature [54]-[59]. In [54], a distributed cooperative secondary control of MG is proposed using feedback linearization. It is assumed that not every DG in the system can directly access the reference frequency and voltage values. By allowing controllers to communicate with their neighbors, they can ensure that frequency and voltage finally reach their reference values. However, the inherently existing coupling between frequency and voltage is not considered. Similarly in [55], the approach is employed to convert the secondary voltage control to a linear second-order tracker synchronization problem.

A general distributed approach has also been proposed in [56] to regulate the frequency, voltage as well as the reactive power. Each DG controller collects all the measurements including frequency, voltage amplitude, and reactive power of other DG units by using a large communication system, then averages them and produces a control signal based on a proportional integral (PI) control scheme. The distributed secondary frequency control laws in [56] are

$$\delta f_{DG_k} = k_{P_f}(f_{MG}^* - \bar{f}_{DG_k}) + k_{i_f}\int(f_{MG}^* - \bar{f}_{DG_k})dt \qquad (1.3)$$

$$\bar{f}_{DG_k} = \frac{\sum_{i=1}^N f_{DG_i}}{N} \qquad (1.4)$$

where δf_{DG_k} is the frequency control output, k_{P_f} and k_{i_f} are the PI controller parameters, f_{MG}^* is the MG frequency reference, and $\bar{f}_{DG_k}$ is the frequency average for all DGs.

Similarly, the distributed secondary voltage control laws in [56] are

$$\delta E_{DG_k} = k_{P_E}(E_{MG}^* - \bar{E}_{DG_k}) + k_{i_E}\int(E_{MG}^* - \bar{E}_{DG_k})dt \qquad (1.5)$$

$$\bar{E}_{DG_k} = \frac{\sum_{i=1}^N E_{DG_i}}{N} \qquad (1.6)$$

where δE_{DG_k} is the voltage control output, k_{P_E} and k_{i_E} are the PI controller parameters, E^*_{MG} is the MG voltage reference, and $\bar{E}_{DG_k}$ is the voltage average for all DGs. However, the distributed control scheme used here requires that each local controller communicates with all the other controllers in the whole system, which almost has the same communication cost as the centralized controller and is quite different from distributed control in multi-agent systems. In addition, no analysis on the dynamics and stability of the whole system is provided.

A distributed averaging PI controller is proposed in [57] to remove the frequency deviation while the power sharing property still holds. A first-order inverter model is derived by applying the results in the theory of coupled oscillators. This is achieved under the assumption that the node voltage remains constant. Also voltage restoration is not considered.

Considering the above-mentioned problems, in Chapter 3, we will consider to restore the frequency to the nominal value while improving the real power sharing accuracy in the secondary control layer.

Voltage quality enhancement is another considered topic in the secondary control hierarchy. In conventional power systems, this issue is usually realized by utilizing the passive voltage compensator which is connected to the transmission line or the voltage transformer [60]. However, in the inverter interfaced MG system, the voltage quality can be directly enhanced by the active compensation method in the DG inverters. Recently, some control approaches have been proposed to control the DG inverter in the islanded MG system to solve the harmonic compensation and unbalanced voltage compensation problem [61]-[65]. The fundamental principle of the harmonic compensation is to make the DG emulate a resistance at harmonic frequencies. While compensating for the voltage unbalance, the approach in the existing literatures is to control the DG in the MG as a negative sequence conductance. The conductance reference is determined by applying a droop function which uses negative sequence reactive power to provide the compensation effort sharing.

In [61], a negative sequence conductance based compensation method is proposed. The proposed method allows even sharing of imbalance current among the DGs and can be integrated with the existing droop control function. A three-phase balancing method using surplus capacity is proposed in [62]. This method controls the inverters to output negative sequence current to compensate for the voltage unbalance within DGs' surplus capacities. In [63], a stationary-frame control method for voltage unbalance compensation in an islanded MG is proposed. This method can make the DGs share the compensation effort autonomously. However, the above methods only deal with the voltage unbalance in the local terminal buses.

To consider the voltage unbalance compensation at some particular bus, e.g., sensitive load bus (SLB), where the loads connected are very sensitive to the unbalanced voltage, a hierarchical control structure for voltage quality enhancement in MG is proposed in [64, 65]. A centralized secondary controller is designed to send proper compensation signals to the primary controllers. However, as pointed out in [66]-[68], there are mainly three limitations with this approach: 1) such a centralized secondary controller is usually costly both in computation and communication when the number of DGs becomes larger and larger; 2) it may suffer from single-point-failures. When a fault occurs in the centralized controller, the whole system may collapse; 3) it is unable to meet the plug-and-play requirement of recent MG system. When some DGs are newly installed or uninstalled, the centralized controller may need to be redesigned [68]. Furthermore, it is worthy to point out that only the unbalanced voltage compensation at a particular location, e.g., SLB, is considered in [64, 65]. This is achieved at the expense of unbalanced voltage outputs of all DGs. To the best of our knowledge, few available results explicitly consider the problem that the terminal outputs of some local buses (LB) are also required to have balanced voltage outputs while ensuring balanced voltage output in SLB.

Similar to the aforementioned discussion, this control strategy has the same disadvantages as the centralized frequency controller. It will be analyzed and modified to a distributed control strategy in Chapter 4.

1.4.3 Tertiary control

The tertiary control is on the top layer of the MG control diagram, and its dynamics response is the slowest one among the three layers. It usually considers the optimal power flow (OPF), economic dispatch (ED), and optimal energy scheduling problem in the MG.

1.4.3.1 Optimal power flow

In the MG control, one of the tough tasks is to carefully control the voltage and prevent abrupt voltage fluctuations, which stem from the well-known sensitivity of voltages to variations of power injections [69]. Hence, the optimal power flow problem becomes more essential in the MG control. Its main objective is to minimize either the power distribution losses or the cost of power drawn from the DGs while effecting voltage regulation.

From the point of view of mathematics, the OPF problem in the MG is deemed challenging because it requires solving nonconvex problems. Conventionally, this problem is solved by applying the methods including the Newton -Raphson method, sequential quadratic optimization, particle swarm optimization, steepest descent-based method, and fuzzy dynamic programming [70]-[73] to obtain a possibly suboptimal solution of the nonconvex problem. However, these methods are computationally cumbersome. Recently, in order to alleviate these drawbacks, a relaxed semidefinite programming (SDP) reformulation of this problem is proposed in [74], which can convert the nonconvex problem to the convex one and achieve a global optimal solution. A balanced distribution system in [75] and an unbalanced distribution system in [76] are modeled to solve the OPF problem, respectively.

1.4.3.2 Economic dispatch

Economic dispatch (ED) is considered one of well-studied and key problems in power system research. It deals with the power allocation among the generators in an economically efficient way while meeting the constraints of total load demand as well as the generator constraints [77]. Some algorithms have been proposed to solve the ED problem, such as quadratic programming [78], λ-iteration [79], Lagrangian relaxation technique [80], and so on. However, all these methods are realized in a centralized way, i.e., collect the global information of all the generators and conduct the optimization in a central node. As pointed out in [81], such a centralized optimization is usually costly both in computation and communication when the power system becomes larger and larger. Moreover, they are unable to meet the plug-and-play requirement of recent smart grid systems. When some generators are newly installed or uninstalled, such centralized optimization may need to be redesigned [82, 83].

Recently in order to overcome the drawbacks mentioned above, distributed algorithms have been proposed [84]-[91]. Their main idea is to decompose the central optimization into several local optimizations. By letting each local optimization agent communicate with its neighbors, the global objective cost function can be minimized. Compared to centralized algorithms, a distributed one has the following major advantages: 1) less computational and communication cost; 2) plug-and-play property required by smart grid systems, which makes algorithm design more flexible; 3) robust to single-point-failures; 4) easy and simple to design and implement as it only handles local information. In [85], ED problem is formulated as the incremental cost λ-consensus problem. The incremental cost of the ith generator λ_i is updated by combining λ_j from its neighbors with the global power mismatch. However, the calculation of the power mismatch term requires the global information of each generator output and the total load demand. A similar λ-consensus algorithm is proposed in [86]. Both incremental cost and power mismatch are obtained in a distributed way through two consensus algorithms. In [87], a distributed ED algorithm with transmission losses is proposed, which is based on two parallel consensus algorithms. An auction-based consensus protocol is proposed in [88].

In addition to λ-consensus, a distributed gradient method has also been applied in the ED problem. In [89], an improved distributed gradient method is proposed to handle both equality and inequality constraints. However, the stepsize should be carefully chosen when the variables reach their constraint bounds. In [90], a fast distributed gradient method consisting of θ-*logarithmic* barrier function is proposed to solve ED problem. Note that the distributed gradient method requires that the initial values should be carefully allocated to meet the equality constraint.

In Chapter 5, we give a distributed economic ED strategy based on projected gradient and finite-time average consensus algorithms for smart grid systems. Besides, two multi-cluster optimization algorithms will be proposed in Chapter 6 to solve the ED problem in a multi-area power system. A new hierarchical decentralized optimization algorithm is also proposed in Chapter 7.

1.4.3.3 *Optimal energy scheduling*

Optimal energy scheduling is a key problem in the electricity market for maintaining the balance between supply and demand [92]. Recently, with the advent of smart grid technologies, a number of information and communication infrastructures have been integrated into existing power systems, which enables real-time communication between the energy provider (supply side) and consumer (demand side) [93, 94]. Thus it offers an additional degree of freedom to do optimal energy management in the demand side. Demand side management (DSM) is envisioned as a key mechanism in the smart grid to effectively reduce the total energy costs and peak to average ratio (PAR) of the total energy demand [95]. The main target of DSM is to alternate the consumer's demand profile in time and/or shape, to make it match the supply [96].

One intuitive method of DSM is called direct load control [97], which requires the utility to acquire free access to the customers. An alternative approach is called smart pricing or real-time pricing (RTP). It uses the electricity price as a measure to manage energy consumption in the demand side. Different from the flat pricing strategy, where a fixed price is announced by the utility for all periods, the price under the RTP strategy is effected by the energy consumption [98]. Numerous DSM research projects have been carried out by applying the RTP strategy [99]-[101]. In [99], a peak load pricing was proposed, where a predetermined distinct price for each period was announced at the beginning. In [100], the RTP was formulated as an optimization problem to maximize the aggregate utility of all the consumers while minimizing the imposed energy cost. A linear and rotational symmetric pricing was proposed in [101], where the price was linearly changed with the variation of the total load demand.

With consideration of RTP, the optimal energy scheduling could be formulated as optimization problems [102]. Such problems are often modeled and analyzed by applying game theory and convex optimization methods [103]-[105]. In [103], a power system consisting of one single utility and several consumers was discussed by using non-cooperative game theory. It is shown that the Nash equilibrium (NE) of this game is the optimal solution of energy cost minimization problem. A multi utilities and multi consumers model was considered in [104], where a Stackelberg game approach was utilized. However, it is noted that the existence of the NE point in a non-cooperative game always highly relies on the game structure as well as the billing policy [105]. Also the NE point is not always a social optimum solution [105]. In other words, if users are behaving in a cooperative way, less cost will be achieved both in individual and global aspects compared to the non-cooperative way. It is thus desired to develop a social optimum solution to the smart grid.

In Chapter 8, a novel real-time pricing strategy will be proposed. An optimal energy scheduling problem is then formulated by minimizing the total social cost of the overall power system. Two different distributed optimization algorithms with different communication strategies are proposed to solve the problem.

1.5 Major Contributions of the Book

The main results of this book are summarized as follows.

1. In Chapter 3, restorations for both voltage and frequency in the droop-controlled inverter-based islanded MG are addressed. A distributed finite-time control approach is used in the voltage restoration which enables the voltages at all

the DGs to converge to the reference value in finite time and thus the voltage and frequency control design can be separated. Then a consensus-based distributed frequency control is proposed for frequency restoration, subject to certain control input constraints. Our control strategies are implemented on the local DGs and thus no central controller is required in contrast to existing control schemes proposed so far. By allowing these controllers to communicate with their neighboring controllers, the proposed control strategy can restore both voltage and frequency to their respective reference values while having accurate real power sharing, under a sufficient local stability condition established. An islanded MG test system consisting of 4 DGs is built in MATLAB$^{\circledR}$ to illustrate our design approach and the results validate our proposed control strategy.

2. In Chapter 4, a distributed cooperative control scheme for voltage unbalance compensation in an islanded MG is presented. By letting each DG share the compensation effort cooperatively, unbalanced voltage in sensitive load bus (SLB) can be compensated. The concept of contribution level for compensation is first proposed for each local DG to indicate its compensation ability. A two-layer secondary compensation architecture consisting of communication layer and compensation layer is designed for each local DG. A totally distributed strategy involving information sharing and exchange is proposed, which is based on finite-time average consensus and newly developed graph discovery algorithm. This strategy does not require the whole system structure as a prior and can detect the structure automatically. The proposed scheme not only achieves similar voltage unbalance compensation performance to the centralized one, but also brings some advantages, such as communication fault tolerance and plug-and-play property. Case studies including communication failure, contribution level variation, and DG plug-and-play are discussed and tested to validate the proposed method.

3. In Chapter 5, we present a distributed economic dispatch strategy based on projected gradient and finite-time average consensus algorithms for smart grid systems. Both conventional thermal generators and wind turbines are taken into account in the economic dispatch model. By decomposing the centralized optimization into optimizations at local agents, a scheme is proposed for each agent to iteratively estimate a solution of the optimization problem in a distributed manner with limited communication among neighbors. It is theoretically shown that the estimated solutions of all the agents reach consensus of the optimal solution asymptomatically. This scheme also brings some advantages, such as plug-and-play property. Different from most existing distributed methods, the private confidential information such as gradient or incremental cost of each generator is not required for the information exchange, which makes more sense in real applications. Besides, the proposed method not only handles quadratic but also non-quadratic convex cost functions with arbitrary initial values. Several case studies implemented on a 6-bus power system as well as an IEEE 30-bus power system are discussed and tested to validate the proposed method.

4. In Chapter 6, we consider an optimization problem which minimizes a global objective function being the sum of local agents' cost functions subject to certain global and local constraints. Besides, both the local cost function and local constraints are only known by the local agent itself. To solve this problem, a novel

distributed algorithm based on projected gradient method is proposed by using synchronous and sequential communication strategies. Due to the large number of agents, the agents are sorted into several groups and each group has a leader to communicate with the leaders of its neighboring groups. A scheme is proposed for each agent to iteratively estimate a solution of the optimization problem in a distributed manner. It is theoretically proved that the estimated solutions of all the agents reach consensus of the optimal solution asymptomatically. Furthermore, this distributed algorithm is applied to solve economic dispatch problems in a multi-area power system. Several case studies implemented on IEEE 30-bus power systems are discussed and tested to validate the proposed method.

5. In Chapter 7, a novel hierarchical decentralized optimization architecture is proposed to solve the economic dispatch problem in a smart grid system. Conventionally such a problem is solved in a centralized way, which is usually costly in computation and not flexible, especially for large-scale power systems. In contrast to centralized algorithms, in this paper we decompose the centralized problem into several local problems. Each local generator only solves its own problem based on its own cost function and generation constraint. An extra coordinator agent is employed to coordinate all the local generator agents. Besides, it also takes responsibility to handle the global demand supply constraint based on a new proposed concept named *virtual agent*. In this way, different from existing distributed algorithms, the global demand supply constraint and local generation constraints are handled separately, which would greatly reduce the computational complexity. It is theoretically shown that under proposed hierarchical decentralized optimization architecture, each local generator agent can obtain the optimal solution in a decentralized fashion. Several case studies implemented on the IEEE 30-bus and the IEEE 118-bus, are discussed and tested to validate the proposed method.

6. In Chapter 8, we propose a novel real-time pricing strategy named proportional and derivative *(PD) pricing*. Conventional real-time pricing strategies only depend on the current total energy consumption. In contrast to it, our proposed pricing strategy also takes the incremental energy consumption into consideration, which aims to further fill the valley load and shave the peak load. An optimal energy scheduling problem is then formulated by minimizing the total social cost of the overall power system. Two different distributed optimization algorithms with different communication strategies are proposed to solve the problem. Several case studies implemented on a heating ventilation and air conditioning (HVAC) system are tested and discussed to show the effectiveness of both the proposed pricing function and distributed optimization algorithms.

1.6　Organization of the Book

The book is mainly organized based on the hierarchical control structure, involving the distributed control and distributed optimization. In Chapters 3 and 4, distributed control scheme is applied to the secondary layer, where a distributed voltage and frequency restoration control method and a distributed voltage unbalance compensation method are proposed, respectively. In Chapters 5 through 8, we develop several distributed optimization algorithms in the tertiary control layer, which includes single-area, multi-area economic dispatch, hierarchical decentralized architecture for ED, and optimal energy scheduling.

Chapter 1 first introduces the motivation and some background, as well as the objectives. Literature review of recent microgrid control strategy is also conducted in Chapter 1, where some control methods based on the hierarchical control strategy are introduced. In Chapter 2, some preliminary knowledge including graph theory, distributed finite-time average consensus (FACA) algorithm, finite-time control theory, and distributed optimization is introduced. In Chapter 3, we propose a distributed secondary control method which aims to restore the voltage and frequency to their nominal values while keeping the real power sharing accuracy. A distributed secondary control scheme for the voltage unbalance compensation is developed in Chapter 4. In Chapter 5, a distributed single-area economic dispatch method is proposed, where both conventional thermal generators and wind turbines are taken into account. Two multi-cluster distributed optimization algorithms are proposed in Chapter 6, where applications to economic dispatch in a multi-area power system are conducted. In Chapter 7, a hierarchical decentralized optimization architecture is proposed to solve the ED problem. In Chapter 8, a distributed optimal energy scheduling problem is considered, where a novel real-time pricing strategy named proportional and derivative (PD) pricing is proposed. Finally, conclusions and suggestions for future research are given in Chapter 9.

Chapter 2

Preliminaries

In this chapter, some basic concepts, definitions and lemmas are introduced including graph theory, distributed finite-time average consensus (FACA) algorithm, finite-time control theory, and distributed optimization. A new algorithm called *graph discovery* is also proposed in this chapter to make FACA be realized in a totally distributed way.

2.1 Graph Theory

A graph is defined as $\mathcal{G} = (\mathcal{V}, \xi)$, where $\mathcal{V} = \{1, \cdots, N\}$ denotes the set of vertices, $\xi \subseteq \mathcal{V} \times \mathcal{V}$ is the set of edges between two distinct vertices. If for all $(i, j) \subseteq \xi$, then $(j, i) \subseteq \xi$, we call $\mathcal{G}$ undirected; otherwise it is called directed graph. The set of neighbors of the ith vertex is denoted as $\mathcal{N}_i \overset{\Delta}{=} \{j \subseteq \mathcal{V} : (i, j) \subseteq \xi\}$. The graph $\mathcal{G}$ is connected, meaning that there exists at least one path between any two distinct vertices. The elements of the adjacency matrix $\mathbf{A}$ are defined as $a_{ij} = a_{ji} = 1$ if $j \subseteq \mathcal{N}_i$; otherwise $a_{ij} = a_{ji} = 0$. In this book, no self communication is allowed, hence $(i, i) \not\subseteq \xi, \forall i$, which implies that $a_{ii} = 0$. The Laplacian matrix of $\mathcal{G}$ is defined as $\mathcal{L} = \Delta - \mathbf{A}$, where Δ is called in-degree matrix and is defined as $\Delta = diag(\Delta_i) \subseteq \mathbb{R}^{N \times N}$ with $\Delta_i = \sum_{j \in \mathcal{N}_i} a_{ij}$. It is well known that the Laplancian matrix $\mathcal{L}$ of an undirected graph has one distinct zero eigenvalue and all the others are positive [106].

2.2 Distributed Finite-Time Average Consensus Algorithm

In distributed multi-agent systems, the average consensus theorem has been widely studied [107]-[109]. It ensures that each agent approaches a consistent understanding of its shared information in a distributed manner. In order to have convergence with finite number of iteration steps, a FACA has been proposed in [110]. Compared to the conventional average consensus algorithms, this algorithm has the following advantages: 1) It can realize consensus in a finite time; 2) It can ensure all the agents reach consensus at the same time.

The general average consensus can be represented as

$$x_i^{l+1} = w_{ii}(l)x_i^l + \sum_{j \in \mathcal{N}_i} w_{ij}(l)x_j^l \qquad (2.1)$$

where x_i^l denotes the information shared by the ith agent at iteration l, w_{ii}, w_{ij} are the update gains of its own states and neighboring states, respectively, $\mathcal{N}_i$ is the set of neighbor agents of the ith agent.

Lemma 2.1 [110] *Let $\lambda_2 \neq \lambda_3 \neq \cdots \neq \lambda_{K+1} \neq 0$ be the K distinct nonzero eigenvalues of the graph Laplacian matrix $\mathcal{L}$, $y_i, i = 1, \cdots, n$ in (2.1) can reach consensus in finite K steps, if the updating gains for agent i are chosen as*

$$w_{ij}(m) = \begin{cases} 1 - \frac{n_i}{\lambda_{m+1}}, & j = i \\ \frac{1}{\lambda_{m+1}}, & j \in \mathcal{N}_i \\ 0, & otherwise \end{cases} , m = 1, \cdots, K \qquad (2.2)$$

where $n_i = |\mathcal{N}_i|$, which is the number of the neighboring agents of agent i.

2.2.1 Distributed FACA

The FACA can achieve consensus in finite steps, which is necessary for our proposed method developed in the next section. However, from (2.2) we know that the main limitation of the FACA is the assumption that each agent needs to know the nonzero eigenvalues of Laplacian matrix $\mathcal{L}$ of the whole communication graph, i.e., the whole graph topology, as a *prior*. This is very restrictive, as in practice each agent does not have the global information of whole graph topology such as the total number of agents N, and the corresponding Laplacian matrix $\mathcal{L}$ at the beginning. In addition, this global information may change due to the addition and removal of certain agents. Clearly this requirement results in the implementation of FACA non-distributed.

To relax this requirement, a new algorithm named *graph discovery* is proposed, which is based on the well-known "network flooding method" proposed in [111]. By applying the proposed algorithm, each agent can determine N and $\mathcal{L}$ by itself automatically. Similar to [112], we only impose the assumption that each agent i has been assigned a unique identifier $ID(i)$, e.g., its IP address.

Algorithm 2.1 (Graph Discovery) *Let $N_i(k)$ denote the neighbor table set obtained by agent $i, i \in \mathcal{V}$ at time step k, which will be determined by the following steps.*

1. *At $k = 0$, each agent $i \in \mathcal{V}$ initializes the table as*

$$N_i(0) = \{ID(i)[ID(j), j \in \mathcal{N}_i]\}$$

   and sends this data to all its neighbors in $\mathcal{N}_i$.

2. *At each step $k \geq 1$, agent i updates its table set $N_i(k)$ as*

$$N_i(k+1) = \bigcup_{j \in \mathcal{N}_i \cup \{i\}} N_j(k)$$

3. *If $N_i(k) = N_i(k-1)$, then agent i stops exchanging information with its neighbors. Otherwise, go to Step 2).*

4. *Let k_f be the first instant at which $N_i(k) = N_i(k-1)$, i.e.,*

$$k_f = \min\{k | N_i(k) = N_i(k-1)\}$$

   then the total number of agents $N = |N_i(k_f)|$, where $|.|$ denotes the number of elements in the table set.

5. *Finally the $N \times N$ Laplacian matrix $\mathcal{L}$ can be extracted from $N_i(k_f)$ according to the definition introduced in Section II-A, e.g., $\mathcal{L}$ can be extracted with the ith row being determined by $N_i(0)$ in $N_i(k_f)$.*

An example with 4 agents to illustrate this algorithm is shown in Table 2.1. For vertices 2 and 3, it takes $k_f = 2$ steps while for vertices 1 and 4, it takes $k_f = 3$ to discover the whole graph.

Note that this algorithm not only discovers the number of agents but also the whole graph topology. By applying Algorithm 2.1 in the initial of the FACA, it can be realized in a totally distributed way.

2.3　Finite-Time Control

The idea of finite-time stabilization is to steer system states to arrive at their equilibrium in finite time. Furthermore, finite-time stable closed-loop systems might have better robustness and disturbance rejection properties [113]. This idea will be used in the voltage restoration control in Chapter 3. To investigate the finite-time stability, some basic concepts and definitions are introduced. For convenience, we denote $sig(x)^\alpha = sgn(x)|x|^\alpha$.

Definition 2.1 [114] *Consider the system*

$$\dot{x} = f(x,t), f(0,t) = 0, x \in \mathbb{U}_0 \subset \mathbb{R}^n \tag{2.3}$$

Table 2.1: Illustration example of Algorithm 2.1

Step	Content	Communication Graph
$k=0$	$G_1(0) = \{ID(1)[ID(2)]\}$ $G_2(0) = \{ID(2)[ID(1),ID(3)]\}$ $G_3(0) = \{ID(3)[ID(2),ID(4)]\}$ $G_4(0) = \{ID(4)[ID(3)]\}$	
$k=1$	$G_1(1) = \{G_1(0),G_2(0)\}$ $G_2(1) = \{G_1(0),G_2(0),G_3(0)\}$ $G_3(1) = \{G_2(0),G_3(0),G_4(0)\}$ $G_4(1) = \{G_3(0),G_4(0)\}$	
$k=2$	$G_1(2) = \{G_1(0),G_2(0),G_3(0)\}$ $G_2(2) = \{G_1(0),G_2(0),G_3(0),G_4(0)\}$ $G_3(2) = \{G_1(0),G_2(0),G_3(0),G_4(0)\}$ $G_4(2) = \{G_2(0),G_3(0),G_4(0)\}$	
$k=3$	$G_1(3) = \{G_1(0),G_2(0),G_3(0),G_4(0)\}$ $G_2(3) = \{G_1(0),G_2(0),G_3(0),G_4(0)\}$ $G_3(3) = \{G_1(0),G_2(0),G_3(0),G_4(0)\}$ $G_4(3) = \{G_1(0),G_2(0),G_3(0),G_4(0)\}$	

where $f : \mathbb{U}_0 \times \mathbb{R}^+ \to \mathbb{R}^n$ is continuous on an open neighborhood $\mathbb{U}_0$ of the origin $x = 0$. The equilibrium $x = 0$ of the system is locally finite-time stable if it is Lyapunov stable and for any initial condition $x(t_0) = x_0 \in \mathbb{U}$ where $\mathbb{U} \subset \mathbb{U}_0$, if there is a setting time $T > t_0$, such that every solution $x(t;t_0,x_0)$ of system (2.3) satisfies $x(t;t_0,x_0) \in \mathbb{U}\backslash\{0\}$ for $t \in [t_0,T)$, and

$$\lim_{t \to T} x(t;t_0,x_0) = 0, x(t;t_0,x_0) = 0, \forall t > T$$

If $\mathbb{U} = \mathbb{R}^n$, then the origin $x = 0$ is a globally finite-time stable equilibrium.

Definition 2.2 [114], [115] *Suppose $(r_1,\cdots,r_n) \in \mathbb{R}^n$ where $r_i > 0$, $i = 1,\cdots,n$, and $V(x_1,\cdots,x_n) : \mathbb{R}^n \to \mathbb{R}$ is a continuous function. Then $V(x_1,\cdots,x_n)$ is homogeneous of degree $\sigma > 0$ with respect to $(r_1,\cdots,r_n)$, if for any given ε,*

$$V(\varepsilon^{r_1} x_1,\cdots,\varepsilon^{r_n} x_n) = \varepsilon^{\sigma} V(x_1,\cdots,x_n)$$

Let $x = (x_1,\cdots,x_n)^T$ and $f(x) = (f_1(x),\cdots,f_n(x))^T$ be a continuous vector field. $f(x)$ is said to be homogeneous of degree $\kappa \in \mathbb{R}$ with respect to $(r_1,\cdots,r_n)$ if for any $\varepsilon > 0$,

$$f_i(\varepsilon^{r_1} x_1,\cdots,\varepsilon^{r_n} x_n) = \varepsilon^{\kappa+r_i} f_i(x), i = 1,\cdots,n, x \in \mathbb{R}^n$$

A given system $\dot{x} = f(x)$ is said to be homogeneous if $f(x)$ is homogeneous.

For a homogeneous system, we have the following lemmas.

Lemma 2.2 [115] *Suppose $\dot{x} = f(x)$ is homogeneous of degree κ. Then the origin of the system is finite-time stable if the origin is asymptotically stable and $\kappa < 0$.*

Lemma 2.3 [116] *For the following system*

$$\dot{x} = y$$
$$\dot{y} = Mu \qquad (2.4)$$

with feedback control input

$$u = -k_1 sig(x)^{\alpha_1} - k_2 sig(y)^{\alpha_2} \qquad (2.5)$$

where $x = \begin{bmatrix} x_1 & \cdots & x_N \end{bmatrix}^T$, $y = \begin{bmatrix} y_1 & \cdots & y_N \end{bmatrix}^T$, $k_1, k_2 > 0$, α_1 *and* α_2 *are two positive constants satisfying* $0 < \alpha_1 < 1$, $\alpha_2 = \frac{2\alpha_1}{1+\alpha_1}$, $M \in \mathbb{R}^{N \times N}$ *is a symmetric positive definite matrix,* $sig(\cdot)$ *is defined element-wisely, it is globally finite-time stable.*

2.4 Multi-Agent Optimization

In this section, some general distributed multi-agent optimization methods are introduced.

Due to potential applications in power system [117, 118], wireless network system [119] and sensor network systems [120], there has been growing research interests in distributed optimization, where local agents cooperatively minimize a global objective cost function which is the sum of local objective functions [121]-[133]. Different from the conventional centralized optimization where information on all the agents is available and collected in one central node, distributed optimization decomposes the central node into several sub-nodes (agents) for reasons including privacy concerns, computational and communication burdens. In distributed optimization, each agent only accesses its local cost function and local constraint set. By letting each local agent communicate with its neighboring agents, the global objective cost function can be minimized. Generally speaking, according to the form of formulated problems, there are mainly two kinds of algorithms to solve distributed optimization, i.e., continuous-time algorithm [121]-[125] and discrete-time algorithm [126]-[133].

The continuous-time algorithm is mainly developed on the basis of well-developed control theory [121]. In [122], a continuous-time proportional-integral distributed optimization method is proposed, where dual decomposition and consensus based method are used. The convergence of continuous-time distributed optimization over directed networks is analyzed in [123]. An event-triggered based continuous-time distributed coordination algorithm is proposed in [124]. However, very few continuous-time algorithms have addressed constrained optimization problems except for [125], where the Karush- Kuhn- Tucker (KKT) condition and Lagrangian multiplier methods are cooperatively used.

Apart from continuous-time algorithms, discrete-time algorithms have also been well studied in distributed optimization. Among them, gradient or subgradient based methods are most popular due to their simplicity and ease of implementation. In [126], two distributed primal-dual subgradient algorithms are developed under inequality and equality constraints, respectively. The convergence rate of the distributed dual subgradient averaging method is analyzed in [127]. A stochastic gradient algorithm is proposed in [128] to solve a distributed non-convex optimization problem. Constrained consensus and constrained optimization problems are considered in [129], where a distributed projected consensus algorithm and a projected gradient method are proposed to solve such problems, respectively.

Besides, recently an interesting distributed optimization algorithm called alternating direction method of multipliers (ADMM) has been proposed and developed [134]. Its main idea is to decompose the original problem into two local subproblems, and then solve them in an alternating fashion [135, 136].

In this book, we develop and implement distributed optimization algorithms mainly based on two general distributed multi-agent optimization methods. According to different communication strategies, we briefly name them as synchronous optimization [130] and sequential optimization [132].

2.4.1 Synchronous optimization

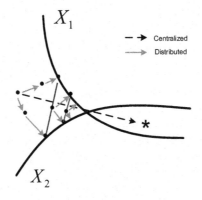

Figure 2.1: Illustration of synchronous optimization

A synchronous multi-agent optimization method is proposed in [130], which mainly consists of two steps. Firstly, all agents update their own estimate based on local information given by the agent's objective function and constraint set. Then, they exchange their estimates by combining the estimates received from its neighbors. This algorithm can be summarized as [130]

$$v^i(k) = \sum_{j=0}^{m} a_j^i(k) x^j(k) \tag{2.6}$$

$$x^i(k+1) = P_{X_i}[v^i(k) - \alpha_k d_i(k)] \tag{2.7}$$

where $a_j^i(k)$ are nonnegative weights, $\alpha_k > 0$ is a stepsize, $d_i(k)$ is a subgradient of local objective function $f_i(x)$ at $x = v^i(k)$, $P_{X_i}[.]$ is a projection operator, which will be introduced later.

The convergence of the algorithm (2.6)-(2.7) with the weight vectors $a_j^i(k) = (1/m, \cdots, 1/m)^T$ is proved. This indicates that $v(k) = v^i(k) = v^j(k), \forall i, j$, which is in fact the average of the estimates from all the agents. The illustration example of this algorithm consisting of 2 agents is shown in Fig. 2.1.

2.4.2 Sequential optimization

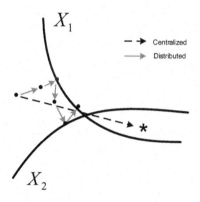

Figure 2.2: Illustration of sequential optimization

Different from synchronous optimization, in sequential optimization the agents sequentially update an iterate sequence in a cyclic or a random order. One well-known sequential optimization algorithm is proposed in [132], which is also known as the incremental subgradient method. Suppose there are m agents in the systems, in one iteration, each agent updates its estimate once. Let x_k be the vector obtained after k iterations, then the vector x_{k+1} is

$$x_{k+1} = \psi_{m,k}, \tag{2.8}$$

where $\psi_{m,k}$ is obtained after m steps

$$\psi_{i,k} = P_{X_i}[\psi_{i-1,k} - \alpha_k g_{i,k}], \quad g_{i,k} \in \partial f_i(\psi_{i-1,k}), \quad i = 1, \cdots, m \tag{2.9}$$

starting with

$$\psi_{0,k} = x_k, \tag{2.10}$$

where $\partial f_i(\psi_{i-1,k})$ denotes the subgradient of f_i at the point $\psi_{i-1,k}$. The illustration example of this algorithm consisting of 2 agents is shown in Fig. 2.2.

2.4.3 Projection

The projection of a vector $\bar{x}$ onto a closed convex set X is defined as

$$P_X[\bar{x}] = arg\min_{x\in X}\|\bar{x}-x\| \tag{2.11}$$

where $\|x\|$ denotes the Euclidean norm, i.e., $\|x\| = \sqrt{x^T x}$, x^T denotes the transpose of vector x.

Two important properties of the projection operation are summarized as follows [131].

(a) **Projection inequality** For any $x \in \mathbb{R}^n$ and all $y \in X$

$$(x - P_X[x])^T (y - P_X[x]) \le 0 \tag{2.12}$$

(b) **Projection non-expansiveness** For any $x, y \in \mathbb{R}^n$

$$\|P_X[x] - P_X[y]\| \le \|x - y\| \tag{2.13}$$

Based on the projection inequality, we have the following lemma.

Lemma 2.4 Let X be a nonempty closed convex set in $\mathbb{R}^n$, then for any $x \in \mathbb{R}^n$ and all $y \in X$,

$$\|x - y\|^2 \ge \|P_X[x] - y\|^2 + \|P_X[x] - x\|^2. \tag{2.14}$$

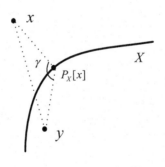

Figure 2.3: Illustration of projection operator

Proof: From (2.12), we have $(x - P_X[x])^T (y - P_X[x]) = \|x - P_X[x]\|\,\|y - P_X[x]\|$ $\cos(\gamma) \le 0$, where γ is the angle between the vectors $x - P_X[x]$ and $y - P_X[x]$ as shown in Fig. 2.3. Thus $\cos(\gamma) \le 0$, i.e.,

$$\cos(\gamma) = \frac{\|P_X[x] - y\|^2 + \|P_X[x] - x\|^2 - \|x - y\|^2}{2\|P_X[x] - y\|\,\|P_X[x] - x\|} \le 0$$

This indicates (2.14) holds. ■

In the following, two basic projection operations are introduced, which will be used in Chapters 5 - 8 later.

2.4.3.1   Projection operator (Case 1)

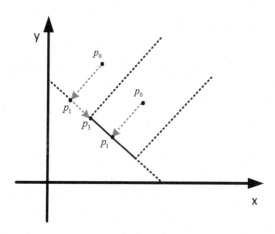

Figure 2.4: Illustration of projection in Case 1 with 2-dimension example

In the formulated economic dispatch problem in Chapters 5 and 6, the constraint set for each agent is convex and is usually in the following form

$$X_i = \left\{ \begin{array}{l} \sum_{i=1}^{N} x_i = C \\ \underline{x}_i \le x_i \le \bar{x}_i \end{array} \right. \tag{2.15}$$

where C is some constant, $\underline{x}_i$ and $\bar{x}_i$ are the lower bound and upper bound of x_i. The projection illustration with a 2-dimension example is shown in Fig. 2.4.

If the inequality constraint is ignored, the constraint set for each agent X_i' is identical, e.g., $\sum_{i=1}^{N} x_i = C$. This can be treated as an "N-dimension" plane in the Hilbert space with the normal vector $\vec{n} = \begin{bmatrix} 1 & \cdots & 1 \end{bmatrix}^T \in \mathbb{R}^{N \times 1}$. The projection operation for a given point $p_0 = [x_1, \cdots, x_N]^T \in \mathbb{R}^{N \times 1}$ to this plane can be easily obtained as

$$P_{X_i'}[p_0] = p_0 - \frac{\vec{n}^T p_0 - P_d}{N} \vec{n}, \quad i = 1, \cdots, N \tag{2.16}$$

Obviously, if $p_0 \in X_i'$, then $P_{X_i'}[p_0] = p_0$.

If the inequality constraint, i.e., $\underline{x}_i \leq x_i \leq \bar{x}_i$ is imposed, then the projection operation should consider the boundary constraint. Let $p_1 = P_{X_{i'}}[p_0]$, if $[p_1]_i > \bar{x}_i$ or $[p_1]_i < \underline{x}_i$, then set $[p_1]_i = \bar{x}_i$ or $[p_1]_i = \underline{x}_i$ respectively, where $[.]_i$ denotes the ith component of the vector. Let $p_2 \in \mathbb{R}^{(N-1)\times 1}$ be the remaining vector by removing $(p_1)_k$, i.e., $p_2 = [x_1, \cdots, x_{i-1}, x_{i+1}, \cdots, x_N]^T$. Then project p_2 onto the new constrain set X_i'', i.e., $\sum_{k=1}^{N} x_k - x_i = C - \bar{x}_i$ or $\sum_{k=1}^{N} x_k - x_i = C - \underline{x}_i$ using the following operations

$$P_{X_{k''}}[p_2] = \begin{cases} p_2 - \dfrac{\vec{n}'^T p_2 - C + \bar{x}_i}{N-1}\vec{n}', & [p_1]_i > \bar{x}_i \\ p_2 - \dfrac{\vec{n}'^T p_2 - C + \underline{x}_i}{N-1}\vec{n}', & [p_1]_i < \underline{x}_i \end{cases} \tag{2.17}$$

where $\vec{n}' = \begin{bmatrix} 1 & \cdots & 1 \end{bmatrix}^T \in \mathbb{R}^{(N-1)\times 1}$. Let $p_3 = P_{X_{i''}}[p_2]$, the final projection result of p_0 with consideration of boundary constraint can be obtained by inserting $(p_1)_i$ into p_3 in the ith place, i.e.,

$$P_{X_k}[p_0] = [(p_3)_1, \cdots, (p_3)_{i-1}, (p_1)_i, (p_3)_i, (p_3)_{N-1}]^T \tag{2.18}$$

2.4.3.2 Projection operator (Case 2)

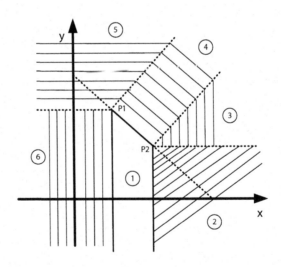

Figure 2.5: Illustration of projection in Case 2 with 2-dimension example

The general form of the constraint set in Chapter 8 is shown as follows

$$M_i = \begin{cases} \underline{x}_i \leq [\mathbf{x}]_i \leq \bar{x}_i \\ \mathbf{1_N}^T \cdot \mathbf{x} \leq 0 \end{cases}, i = 1, \cdots, N \tag{2.19}$$

Then the projection operation for a given point $p_0 = [x_1, x_2, \cdots, x_N]^T$ can be easily divided into 6 cases, which is shown in Fig. 2.5 with a 2-dimension example ($N=2$).

1) Case 1: $p_0 \in M_i$, then the projection is $P_{M_i}[p_0] = p_0$.

2) Case 2: $p_0 \in \{x \in \mathbb{R}^{(N)\times 1} | x_i > \bar{x}_i, \sum_{k=1}^{N} x_k - x_i < -\bar{x}_i\}$, then the projection is to project p_0 into the plane $x_i = \bar{x}_i$, i.e.,

$$P_{M_i}[p_0] = \begin{bmatrix} x_1 & \cdots & x_{i-1} & \bar{x}_i & x_{i+1} & \cdots & x_N \end{bmatrix}^T$$

3) Case 3: $p_0 \in \{x \in \mathbb{R}^{(N)\times 1} | -\bar{x}_i \le \sum_{k=1}^{N} x_k - x_i \le x_i - 2\bar{x}_i\}$, then the projection is to project p_0 into the edge, which can be summarized as follows: Let $p_1 \in \mathbb{R}^{N-1\times 1}$ be the resulting vector after removing $[p_0]_i$, where $[.]_i$ denotes the ith component of the vector, i.e., $p_1 = [x_1, \cdots, x_{i-1}, x_{i+1}, \cdots, x_N]^T$. Then project p_1 onto a new constrain set M_i', i.e., $\sum_{k=1}^{N} x_k - x_i = -\bar{x}_i$ using the following operation

$$P_{M_i'}[p_1] = p_1 - \frac{\vec{n}'^T p_1 + \bar{x}_i}{N}\vec{n}' \tag{2.20}$$

where $\vec{n}' = \begin{bmatrix} 1 & \cdots & 1 \end{bmatrix}^T \in \mathbb{R}^{N-1\times 1}$. Letting $p_2 = P_{M_i'}[p_1]$, the final projection result of p_0 with consideration of boundary constraint can be obtained by inserting $\bar{x}_i$ into p_2 in the ith place, i.e.,

$$P_{M_i}[p_0] = [[p_2]_1, \cdots, [p_2]_{i-1}, \bar{x}_i, [p_2]_i, [p_2]_N]^T \tag{2.21}$$

4) Case 4: $p_0 \in \{x \in \mathbb{R}^{(N)\times 1} | \sum_{k=1}^{N} x_k - x_i > x_i - 2\bar{x}_i, \sum_{k=1}^{N+1} x_k - x_i < x_i - 2\underline{x}_i, \sum_{k=1}^{N} x_k - x_i > -x_i\}$, then the projection is to project p_0 onto the N-dimension plane $\sum_{i=1}^{N} x_i = 0$, i.e.,

$$P_{M_i}[p_0] = p_0 - \frac{\vec{n}^T p_0}{N}\vec{n} \tag{2.22}$$

where $\vec{n} = \begin{bmatrix} 1 & \cdots & 1 \end{bmatrix}^T \in \mathbb{R}^{(N)\times 1}$.

5) Case 5: $p_0 \in \{x \in \mathbb{R}^{(N)\times 1} | \sum_{j=1}^{N} x_j - x_i > x_i - 2\underline{x}_i, \sum_{k=1}^{N} x_k - x_i \ge -\underline{x}_i\}$, the projection in this case is similar to Case 3 except replacing $\bar{x}_i$ in (2.20), (2.21) by $\underline{x}_i$.

6) Case 6: $p_0 \in \{x \in \mathbb{R}^{(N)\times 1} | x_i < \underline{x}_i, \sum_{k=1}^{N} x_k - x_i < -\underline{x}_i\}$, the projection in this case is similar to Case 2, which is to project p_0 onto the plane $x_i = \underline{x}_i$, i.e.,

$$P_{M_i}[p_0] = \begin{bmatrix} x_1 & \cdots & x_{i-1} & \underline{x}_i & x_{i+1} & \cdots & x_N \end{bmatrix}^T$$

DISTRIBUTED
SECONDARY
CONTROL

Chapter 3

Distributed Voltage and Frequency Restoration Control

In this chapter, we start to present the main results of this book by designing distributed controllers for voltage and frequency restoration in the secondary layer of the droop-controlled inverter-based islanded MG.

A distributed finite-time control approach is used in the voltage restoration which enables the voltages at all the DGs to converge to the reference value in finite time and thus the voltage and frequency control design can be separated. Then a consensus-based distributed frequency control is proposed for frequency restoration, subject to certain control input constraints. Our control strategies are implemented on the local DGs and thus no central controller is required in contrast to existing control schemes proposed so far. By allowing these controllers to communicate with their neighboring controllers, the proposed control strategy can restore both voltage and frequency to their respective reference values while having accurate real power sharing, under a sufficient local stability condition established. An islanded MG test system consisting of 4 DGs is built in MATLAB$^{\circledR}$ to illustrate our design approach and the results validate our proposed control strategy.

3.1 Introduction

In this chapter, we focus on the secondary control layer, where a distributed control structure for an islanded MG system is to be proposed.

We start with analyzing the dynamics of the DG units and the power network. Under the assumption of a purely inductive transmission line, a simplified dynamic model of the islanded MG is derived. We next design distributed voltage controllers to restore the voltage magnitudes to their reference values for all DGs within finite time. Then a consensus-based distributed frequency control is designed. In the case of frequency restoration, there also exists a challenge that the control inputs should be equal to each other in their steady state in order to meet the power sharing property. A distributed proportional and integral method is proposed to handle such a constraint. A sufficient condition for frequency restoration is also established.

In summary, the main contributions of this chapter are three-fold.

1. Both frequency and voltage restoration are addressed. A finite-time voltage control is first proposed in islanded MG control to ensure that the voltage magnitudes are restored to their reference values for all DGs within finite time, regardless of frequency control. This enables the frequency and voltage control to be designed separately.

2. Distributed consensus-based frequency control method is derived to restore the frequency to its reference value while maintaining the real power sharing accuracy.

3. In contrast to existing schemes that require a central computing and communication controller, the proposed distributed secondary control strategy is implemented on local DG controllers, which can avoid single-point failure and thus is more reliable and economically efficient.

3.2   Modeling of MG

A schematic diagram of a generic islanded MG with a total of N DGs is shown in Fig. 3.1. Each DG has been connected with respective local loads. They are integrated through an MG network. Hence the model of an islanded MG mainly consists of two parts, i.e., DG model and MG network model.

3.2.1   DG model

In a general AC MG system, each DG consists of a prime DC source, a DC/AC inverter, an LCL filter, and an RL output connector [29], as shown in Fig. 3.2. These inverters operate in the voltage control mode when the MG is islanded [26]. Generally, there are three control loops, *namely*, voltage control loop, current control loop, and droop control loop, in the primary DG controller. Its detailed mathematical model is introduced in [137]. It is also analyzed in [137] that there is a wide frequency range for this primary controller. The dynamics of the LCL filter, RL output connector and the voltage and current control loop are much faster than that of the droop control function. Hence we can remodel the primary controller by only considering the dynamics of the droop control function and neglecting the dynamics of the other four fast-dynamic blocks.

The droop control is to regulate the phase angle δ_i and voltage amplitude V_i by

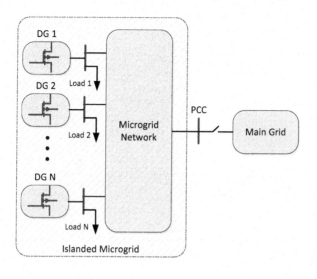

Figure 3.1: Schematic diagram of a generic islanded MG

using the locally measured real power and reactive power information, respectively. According to the CERTS droop control function [10], the phase angle and voltage droop of the ith DG are

$$\dot{\delta}_i = \omega^d - k_{P_i}(P_i^m - P_i^d) \tag{3.1}$$

$$k_{V_i}\dot{V}_i = (V^d - V_i) - k_{Q_i}(Q_i^m - Q_i^d) \tag{3.2}$$

where ω^d, V^d are the desired frequency and voltage amplitude, respectively, k_{V_i} is voltage control gain, k_{P_i} and k_{Q_i} are the frequency and voltage droop gain, respectively, P_i^m and Q_i^m are the measured real and reactive powers, P_i^d and Q_i^d are the desired real and reactive powers, respectively.

The measured P_i^m and Q_i^m can be obtained through the following first-order low-pass filters as

$$\tau_{P_i}\dot{P}_i^m = -P_i^m + P_i \tag{3.3}$$

$$\tau_{Q_i}\dot{Q}_i^m = -Q_i^m + Q_i \tag{3.4}$$

where τ_{P_i} and τ_{Q_i} denote the respective time constants of the two filters, P_i and Q_i are the real and reactive power output of the ith DG.

The derivative of the phase angle can be represented in terms of ω_i as follows:

$$\dot{\delta}_i = \omega_i \tag{3.5}$$

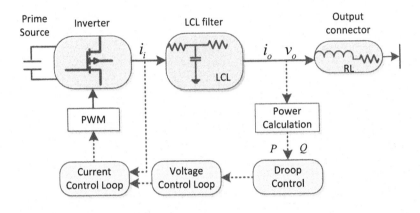

Figure 3.2: The scheme of a local primary inverter controller

Substituting (3.3)-(3.4) to (3.1), (3.2) and (3.5) yields

$$\tau_{P_i}\dot{\omega}_i + \omega_i - \omega^d + k_{P_i}(P_i - P_i^d) = 0 \tag{3.6}$$

$$\tau_{Q_i}k_{V_i}\ddot{V}_i + (\tau_{Q_i} + k_{V_i})\dot{V}_i + V_i - V^d + k_{Q_i}(Q_i - Q_i^d) = 0 \tag{3.7}$$

Equations (3.6)-(3.7) denote a simplified model of the ith DG.

3.2.2 Network model

An MG distribution network is considered as a connected and complex-weighted graph $\mathcal{G} = (\mathcal{V}, \xi)$ with nodes $\mathcal{V}$ being the DGs (buses) and edges ξ being the line impedances. Consider a network with N DGs and let Y_{ik} be the admittance between the ith and k^{th} DG, which is defined as $Y_{ik} = G_{ik} + jB_{ik} \in \mathbb{C}$, where $G_{ik} \in \mathbb{R}$ and $B_{ik} \in \mathbb{R}$ are the conductance and susceptance, respectively. If there is no connection between the ith and k^{th} DG, we define $Y_{ik} = 0$. The set of neighboring DGs of the ith DG is defined as $\mathcal{N}_i = \{k | k \in N, k \neq i, Y_{ik} \neq 0\}$. We also define $G_{ii} = \sum_{k \in \mathcal{N}_i} G_{ik}$, $B_{ii} = \sum_{k \in \mathcal{N}_i} B_{ik}$. We assume that local loads are connected to each DGs, i.e., $S_{L_i} = P_{L_i} + Q_{L_i}$. To incorporate various types of loads, the ZIP load model is applied here [138], which is expressed as

$$P_{L_i} = P_{1_i}V_i^2 + P_{2_i}V_i + P_{3_i} \tag{3.8}$$

$$Q_{L_i} = Q_{1_i}V_i^2 + Q_{2_i}V_i + Q_{3_i} \tag{3.9}$$

where P_{1_i}, Q_{1_i} are nominal constant impedance loads, P_{2_i}, Q_{2_i} are nominal constant current loads, P_{3_i}, Q_{3_i} are nominal constant power loads.

Based on power balance relations [139], the injected real power $\hat{P}_i$ and reactive power $\hat{Q}_i$ are obtained as

$$\hat{P}_i = V_i^2 G_{ii} - \sum_{k \in \mathcal{N}_i} V_i V_k |Y_{ik}| \cos(\delta_i - \delta_k - \phi_{ik}) \qquad (3.10)$$

$$\hat{Q}_i = -V_i^2 B_{ii} - \sum_{k \in \mathcal{N}_i} V_i V_k |Y_{ik}| \sin(\delta_i - \delta_k - \phi_{ik}) \qquad (3.11)$$

where V_i and δ_i are the voltage magnitude and the phase angle of the ith DG, $|Y_{ik}|$ is the magnitude of the admittance Y_{ik}, i.e., $|Y_{ik}| = \sqrt{G_{ik}^2 + B_{ik}^2}$, ϕ_{ik} is the admittance angle of Y_{ik}, i.e., $\phi_{ik} = \phi_{ki} = \arctan(B_{ik}/G_{ik})$. The real and reactive power outputs of the ith DG are given as

$$P_i = P_{L_i} + \hat{P}_i \qquad (3.12)$$

$$Q_i = Q_{L_i} + \hat{Q}_i \qquad (3.13)$$

Combining (3.6)-(3.13), we can get the whole system dynamics.

In order to formulate the voltage and frequency restoration control proposed in the next section, we make the following assumption.

Assumption 3.1 *The power transmission lines of the MG network are lossless, that is, $G_{ik} = 0$, $Y_{ik} = jB_{ik}$, $\phi_{ik} = \phi_{ki} = -\frac{\pi}{2}$, $\forall i \in N, k \in \mathcal{N}_i$.*

Under Assumption 3.1, (3.12)-(3.13) can be reduced to

$$P_i = P_{L_i} + \sum_{k \in \mathcal{N}_i} V_i V_k |B_{ik}| \sin(\delta_i - \delta_k) \qquad (3.14)$$

$$Q_i = Q_{L_i} + V_i^2 \sum_{k \in \mathcal{N}_i} |B_{ik}| - \sum_{k \in \mathcal{N}_i} V_i V_k |B_{ik}| \cos(\delta_i - \delta_k) \qquad (3.15)$$

Remark 3.1 *The assumptions made above are common and reasonable in power system analysis. The purely inductive transmission lines can be achieved by making the inverter output admittance be inductive and dominate over any resistive effects in the network. More detailed justification can be seen in [139], [140].*

Remark 3.2 *Combining (3.6)-(3.15), we conclude that the voltage dynamic (3.7) and (3.15) and the frequency dynamic (3.6) and (3.14) affect each other. In order to design the voltage and frequency control separately, we first control the voltage to its reference value within finite time and then design the frequency restoration control law with constant voltage amplitudes, which will be introduced in the next section.*

3.3 Distributed Secondary Controller Design

In this section, we first analyze some properties of the primary control by studying the behaviors of the node states including both voltage and frequency. Based on these, we propose our secondary control objectives and at last design a distributed secondary control law for the MG to achieve these objectives.

3.3.1 Control objective

From related work in [57] and [141], we conclude that the frequency of the MG under the primary controller with the droop function will be synchronized. According to [142], the synchronized steady-state frequency is $\omega_{ss} = \omega^d + \dfrac{\sum\limits_{i=1}^{N}(P_i^d - P_i)}{\sum\limits_{i=1}^{N}\frac{1}{k_{P_i}}}$. Clearly this shows that as long as the total nominal power output $\sum\limits_{i=1}^{N} P_i^d$ is different from the total consumption real power $\sum\limits_{i=1}^{N} P_i$, the synchronized frequency will deviate from the nominal frequency ω^d. As the frequency is a global state, the droop control function can share real power precisely with the inverse of the droop gain, i.e., $k_{P_i}(P_i - P_i^d) = k_{P_k}(P_k - P_k^d)$, $\forall i, k \in N$. Similarly, the voltage will deviate from its nominal value if the reactive power Q_i is different from its desired value Q_i^d. Thus, secondary control laws need to be designed for both frequency and voltage restoration.

In this chapter, our control objectives for islanded MG are

1. Restore the network frequency and voltage to their respective reference values, i.e.,

$$\lim_{t \to \infty} \omega_i(t) = \omega^{ref}, \forall i \in N$$

$$\lim_{t \to T} V_i(t) = V^{ref}, V_i(t) = V^{ref}, \forall t > T, \forall i \in N$$

for some finite-time T.

2. Guarantee real power sharing accuracy, i.e.,

$$\frac{P_i}{P_k} = \frac{m_i}{m_k}, \forall i, k \in N \tag{3.16}$$

where $m_i, m_k \in \mathbb{R}^+$ are the real power sharing gains and are chosen according to the power rating of the DGs [57]. Conventionally, in the primary control the frequency droop gain k_{P_i} are usually chosen as the inverse proportion of the power rating. Hence we also have (3.16) as

$$\frac{P_i}{P_k} = \frac{m_i}{m_k} = \frac{k_{P_k}}{k_{P_i}}, \forall i, k \in N \tag{3.17}$$

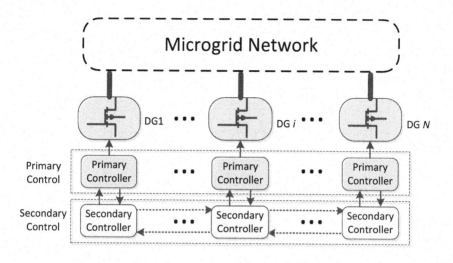

Figure 3.3: Distributed secondary control diagram of an islanded MG

3.3.2 Distributed secondary controller design

A distributed control strategy is proposed in the secondary control layer, as shown in Fig. 3.3. Different from the MGCC control strategy, our secondary controller is applied locally with communication with its neighboring controllers. We add the secondary control inputs $\mathbf{u}_i = \begin{bmatrix} u_i^\omega & u_i^V \end{bmatrix}^T$ into the primary control model (3.6)-(3.7) in the following form:

$$\tau_{P_i} \dot{\omega}_i + \omega_i - \omega^d + k_{P_i}(P_i - P_i^d) + u_i^\omega = 0 \tag{3.18}$$

$$\tau_{Q_i} k_{V_i} \ddot{V}_i + (\tau_{Q_i} + k_{V_i})\dot{V}_i + V_i - V^d + k_{Q_i}(Q_i - Q_i^d) + u_i^V = 0 \tag{3.19}$$

where u_i^ω and u_i^V are the secondary frequency and voltage control input, respectively.

3.3.2.1 Finite-time voltage restoration

The voltage dynamic (3.19) can be rewritten as the following second-order nonlinear system

$$\dot{x}_i = f_i(x_i, x_k) + g_i(x_i)u_i, i \in N, k \in \mathcal{N}_i \tag{3.20}$$

where $x_i = \begin{bmatrix} V_i & \dot{V}_i \end{bmatrix}^T$, $f_i(x_i, x_k) = \begin{bmatrix} \dot{V}_i & f_{1i}(x_i, x_k) \end{bmatrix}^T$, $g_i(x_i) = \begin{bmatrix} 0 & \frac{1}{\tau_{Q_i} k_{V_i}} \end{bmatrix}^T$, $u_i = u_i^V$, with

$$
\begin{aligned}
f_{1i}(x_i, x_k) = &-\frac{\tau_{Q_i} + k_{V_i}}{\tau_{Q_i} k_{V_i}} \dot{V}_i - \frac{k_{Q_i}\left(Q_{1i} + \sum\limits_{k \in N_i} |B_{ik}|\right)}{\tau_{Q_i} k_{V_i}} V_i^2 \\
&+ \frac{k_{Q_i}}{\tau_{Q_i} k_{V_i}} \sum_{k \in N_i} |B_{ik}| V_i V_k \cos(\delta_i - \delta_k) \\
&- \frac{1 + k_{Q_i} Q_{2i}}{\tau_{Q_i} k_{V_i}} V_i - \frac{k_{Q_i}(Q_{3i} - Q_i^d) - V^d}{\tau_{Q_i} k_{V_i}}.
\end{aligned}
$$

Similar to [54], it is assumed that only one DG has access to the reference voltage value, hence we define the voltage local neighborhood tracking error as

$$
e_i^V = \sum_{k \in \mathcal{N}_{C_i}} (V_i - V_k) + g_i^V \left(V_i - V^{ref}\right) \tag{3.21}
$$

$$
e_i^{dV} = \sum_{k \in \mathcal{N}_{C_i}} \left(\dot{V}_i - \dot{V}_k\right) + g_i^V \left(\dot{V}_i - 0\right) \tag{3.22}
$$

where g_i^V is the voltage pinning gain, which is nonzero for the DG that has direct access to the reference voltage value V^{ref}. $\mathcal{N}_{C_i}$ indicates the communication neighborhood set of the ith controller.

We select

$$
y = h_i(x_i) = V_i - V^{ref} \tag{3.23}
$$

Then by using the following coordinate transformation [143]

$$
\begin{cases}
z_{1_i} = h_i(x_i) = V_i - V^{ref} \\
z_{2_i} = L_{f_i} h_i(x_i) = \dot{V}_i
\end{cases} \tag{3.24}
$$

system (3.20) is rewritten as

$$
\begin{cases}
\dot{z}_{1_i} = z_{2_i} \\
\dot{z}_{2_i} = v_i
\end{cases} \tag{3.25}
$$

where

$$
v_i = L_{f_i}^2 h_i(x_i) + L_{g_i} L_{f_i} h_i(x_i) u_i. \tag{3.26}
$$

$L_f h(x)$ is the Lie derivative of $h(x)$ with respect to f, which is defined as $L_f h(x) = \nabla h(x) \cdot f$. Also $L_f^2 h(x)$ is defined as $L_f^2 h(x) = L_f(L_f h(x))$. Note that both $h(x)$ and f_i are continuous and differentiable in their domain. Hence Lie derivative $L_f h(x)$ and $L_f^2 h(x)$ always exist [144].

For the system (3.25), we can construct a distributed finite-time controller as

$$v_i = -k_1 sig(e_i^V)^{\alpha_1} - k_2 sig(e_i^{dV})^{\alpha_2} \tag{3.27}$$

where $k_1, k_2 > 0, 0 < \alpha_1 < 1, \alpha_2 = \frac{2\alpha_1}{1+\alpha_1}$.

From (3.24), (3.26) and (3.27), we can obtain a finite-time voltage controller for (3.19) as

$$u_i^V = u_i = -\frac{k_1 sig(e_i^V)^{\alpha_1} + k_2 sig(e_i^{dV})^{\alpha_2} + L_{f_i}^2 h_i(x_i)}{L_{g_i} L_{f_i} h_i(x_i)} \tag{3.28}$$

where $L_{f_i}^2 h_i(x_i) = f_{1_i}(x_i), L_{g_i} L_{f_i} h_i(x_i) = \frac{1}{\tau_{Q_i} k_{V_i}}$.

Theorem 3.1 *The voltage dynamics system (3.19) with the distributed voltage control (3.28) is globally finite-time stable and can restore the voltages at all the DGs to the reference value in finite time.*

Proof: The proof of this theorem is equivalent to show that system (3.25) with the distributed control input (3.26) is globally finite-time stable. The overall system of (3.25) and (3.26) can be written as

$$\begin{cases} \dot{z}_1 = z_2 \\ \dot{z}_2 = -k_1 sig((\mathcal{L}_c + G^V)z_1)^{\alpha_1} - k_2 sig((\mathcal{L}_c + G^V)z_2)^{\alpha_2} \end{cases} \tag{3.29}$$

where $z_1 = \begin{bmatrix} z_{1_1} & \cdots & z_{1_N} \end{bmatrix}^T, z_2 = \begin{bmatrix} z_{2_1} & \cdots & z_{2_N} \end{bmatrix}^T, \mathcal{L}_c$ is the Laplacian matrix of the designed communication graph, $G^V = diag(g_1^V, g_2^V, \cdots, g_N^V)$.

Let $M = \mathcal{L}_c + G^V$, and clearly it is a symmetric positive definite matrix [145]. Let $x = M z_1, y = M z_2$, then (3.29) becomes (2.4) with (2.5). According to Lemma 2.3, system (3.29) is globally finite-time stable, so Theorem 3.1 holds. ■

Remark 3.3 *The proposed finite-time voltage controller (3.28) can steer the voltage amplitudes to their reference values within finite time T. This is achieved independent of the frequency variable which means Theorem 3.1 is valid no matter what frequency control law is used. This enables the voltage and frequency design to be separated.*

3.3.2.2 Frequency restoration

In order to restore the frequency to the nominal value while guaranteeing the real power sharing accuracy (3.17), from (3.17) and (3.18) we can easily conclude that there are constraints on the control inputs u_i^ω, i.e.,

$$(u_i^\omega)_s/(u_k^\omega)_s = 1, \forall i, k \in N \tag{3.30}$$

where $(u_i^\omega)_s$ indicates the ith frequency control input value in the steady state. Thus (3.30) requires the control inputs in the steady state should be equal to each other.

In order to achieve the control objectives subject to input constraints in (3.30), a distributed proportional and integral method is proposed here. Motivated by the work in [146], the frequency control input is designed as

$$u_i^\omega = \alpha_i(\hat{\omega}_i - \omega_i) \tag{3.31}$$

$$\dot{\omega}_i = \beta_i e_i{}^\omega + \gamma_i \left(\sum_{k \in \mathcal{N}_{C_i}} \left(u_k{}^\omega - u_i{}^\omega \right) \right) \qquad (3.32)$$

$$e_i{}^\omega = \sum_{k \in \mathcal{N}_{C_i}} \left(\omega_i - \omega_k \right) + g_i{}^\omega \left(\omega_i - \omega^{ref} \right) \qquad (3.33)$$

where $\alpha_i, \beta_i, \gamma_i \in \mathbb{R}^+$ are the proportional gains, $e_i{}^\omega$ is defined as frequency local neighborhood tracking error, $g_i{}^\omega$ is the frequency pinning gain, which is nonzero for the DG that has direct access to the reference frequency value ω^{ref}.

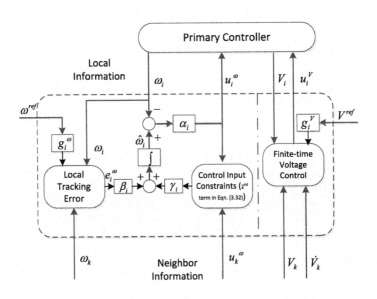

Figure 3.4: The diagram of the proposed secondary controller

A block diagram of a proposed secondary controller is shown in Fig. 3.4. From (3.31)-(3.33) and Fig. 3.4, we can see that the secondary frequency control input $u_i{}^\omega$ contains two parts. The first one, *namely* a local tracking error, is to make the steady-state frequency track the reference frequency, i.e., $\lim_{t \to \infty} \omega_i(t) = \omega^{ref}, \forall i \in N$. The second part is to ensure that the steady-state control input constraints in (3.30) are satisfied.

Note that some interesting results on consensus of multi-agent systems constrained by nonlinear transmissions have been established in [147]. In our case, it is sufficient to allow linear transmission, because in a practical microgrid communication system, the accuracy of data transmission can always be guaranteed by linear transmission. In the sense that the nonlinear function f_{ij} in [147] is set to be

$f_{ij}(x) = x$, our designed consensus protocol (3.31)-(3.33) is consistent with that in [147].

We now analyze the proposed distributed frequency restoration controller (3.31)-(3.33) and establish a sufficient condition for system stability. For the resulting closed loop system, finite time escape will not occur. So in the time duration $\begin{bmatrix} t_0 & t_0 + T \end{bmatrix}$, no signal of the frequency control system will go to infinity. Thus we only need to check the stability of the proposed frequency control after $t > t_0 + T$.

Note that without loss of generality, here we set $P_i^d = 0$, which will not affect the stability analysis.

The proposed distributed secondary controller (3.31)-(3.33) of the whole MG system can be written in the following compact form:

$$U^\omega = \alpha(\hat{\omega} - \omega) \tag{3.34}$$

$$\dot{\hat{\omega}} = \beta e^\omega - \gamma \mathcal{L}_c U^\omega \tag{3.35}$$

$$e^\omega = (\mathcal{L}_c + G^\omega)(\omega - \omega^{ref} \mathbf{1}_{N \times 1}) \tag{3.36}$$

where $U^\omega = \begin{bmatrix} u_1^\omega & u_2^\omega & \cdots & u_N^\omega \end{bmatrix}^T$, $\hat{\omega} = \begin{bmatrix} \hat{\omega}_1 & \hat{\omega}_2 & \cdots & \hat{\omega}_N \end{bmatrix}^T$, $\omega = \begin{bmatrix} \omega_1 & \omega_2 & \cdots & \omega_N \end{bmatrix}^T$, $\alpha = diag(\alpha_1, \alpha_2, \cdots, \alpha_N)$, $\beta = diag(\beta_1, \beta_2, \cdots \beta_N)$, $e^\omega = \begin{bmatrix} e_1^\omega & e_2^\omega & \cdots & e_N^\omega \end{bmatrix}^T$, $\gamma = diag(\gamma_1, \gamma_2, \cdots, \gamma_N)$, $G^\omega = diag(g_1^\omega, g_2^\omega, \cdots, g_N^\omega)$, $\mathcal{L}_c$ is the Laplacian matrix of the communication graph, $\mathbf{1}_{N \times 1}$ denotes the N-dimension vector with the elements all equal to 1.

Similarly, a compact form of (3.18) is as follows:

$$\tau_P^{-1} \dot{\omega} = -\omega + \omega^d \mathbf{1}_{N \times 1} - K_P P(\delta) - U^\omega \tag{3.37}$$

where $\tau_P = diag(\tau_{P_1}^{-1}, \tau_{P_2}^{-1}, \cdots, \tau_{P_N}^{-1})$, $K_P = diag(K_{P_1}, K_{P_2}, \cdots, K_{P_N})$, $P(\delta) = diag(P_1(\delta), P_2(\delta), \cdots, P_N(\delta))$ with V_i being replaced by their reference value V^{ref}.

Combining (3.34)-(3.37), we can obtain the state-space equation as

$$\begin{cases} \dot{\hat{\omega}} = (\beta W + \gamma \mathcal{L}_c \alpha)\omega - \gamma \mathcal{L}_c \alpha \hat{\omega} - \beta W \omega^{ref} \mathbf{1}_{N \times 1} \\ \dot{\delta} = \omega \\ \dot{\omega} = \tau_P(\alpha - I_N)\omega - \tau_P \alpha \hat{\omega} - \tau_P K_P P(\delta) + \tau_P \omega^d \mathbf{1}_{N \times 1} \end{cases} \tag{3.38}$$

where $W = \mathcal{L}_c + G^\omega$.

Theorem 3.2 *Consider the closed-loop frequency control system (3.38). The proposed distributed controller (3.31)-(3.33) ensures that i) system (3.38) is locally exponentially stable; ii) the frequency converges to its reference value, i.e.,* $\lim_{t \to \infty} \omega_i(t) = \omega^{ref}, \forall i \in N$ *while the input constraints in (3.30) are satisfied; iii) The power sharing accuracy in (3.17) is guaranteed if matrix A has exactly one zero eigenvalue and all its other eigenvalues are in the open left half complex plane, where*

$$A = \begin{bmatrix} -\gamma \mathcal{L}_c \alpha & \mathbf{0}_{N \times N} & \gamma \mathcal{L}_c \alpha + \beta W \\ \mathbf{0}_{N \times N} & \mathbf{0}_{N \times N} & I_N \\ -\tau_P \alpha & -\tau_P K_P \mathcal{L}_w & \tau_P(\alpha - I_N) \end{bmatrix} \tag{3.39}$$

with $\mathcal{L}_w$ being a weighted Laplacian matrix which is defined as

$$\mathcal{L}_w = \mathbf{B}diag(V^{ref}V^{ref}|B_{ik}|\cos(\delta_i^\star - \delta_j^\star))\mathbf{B}^T,$$

where $\mathbf{B}$ is the incidence matrix of the communication graph, $\delta_i^\star$ is the voltage angle of the ith DG at the equilibrium point.

Proof : Define the frequency error as $\tilde{\omega} = \omega - \omega^{ref}\mathbf{1}_{N \times 1}$, then $\tilde{\delta}(t) = \delta(0) + \int_0^t \tilde{\omega}(\tau)d\tau$. Note that the dynamics (3.38) is not dependent on the value of the angles δ_i, but only on their differences $\delta_i - \delta_j$. Thus, we can arbitrarily choose one node, say node N, as a reference node and express $\tilde{\delta}_i, i = 1, 2, \cdots, N-1$ relative to $\tilde{\delta}_N$ via the state transformation

$$\theta = \mathcal{R}\tilde{\delta}, \quad \mathcal{R} = \begin{bmatrix} I_{N-1} & -\mathbf{1}_{(N-1) \times 1} \end{bmatrix} \tag{3.40}$$

where $\theta = \begin{bmatrix} \theta_1 & \theta_2 & \cdots & \theta_{N-1} \end{bmatrix}^T$ is a $(N-1) \times 1$ vector. Then we can express the system in a new coordinate as

$$\begin{cases} \dot{\hat{\omega}} = (\beta W + \gamma \mathcal{L}_c \alpha)\tilde{\omega} - \gamma \mathcal{L}_c \alpha \hat{\omega} + \gamma \mathcal{L}_c \alpha \omega^{ref}\mathbf{1}_{N \times 1} \\ \dot{\theta} = \mathcal{R}\tilde{\omega} \\ \dot{\tilde{\omega}} = \tau_P(\alpha - I_N)\tilde{\omega} - \tau_P \alpha \hat{\omega} - \tau_P K_P P(\theta) \\ \quad\quad + \tau_P(\omega^d + (\alpha - I_N)\omega^{ref})\mathbf{1}_{N \times 1} \end{cases} \tag{3.41}$$

Let $(\hat{\omega}^\star, \theta^\star, \tilde{\omega}^\star)$ be the equilibrium of (3.41), which satisfies

$$\begin{cases} (\beta W + \gamma \mathcal{L}_c \alpha)\tilde{\omega}^\star - \gamma \mathcal{L}_c \alpha \hat{\omega}^\star + \gamma \mathcal{L}_c \alpha \omega^{ref}\mathbf{1}_{N \times 1} = \mathbf{0}_{N \times 1} \\ \mathcal{R}\tilde{\omega}^\star = \mathbf{0}_{(N-1) \times 1} \\ \tau_P(\alpha - I_N)\tilde{\omega}^\star - \tau_P \alpha \hat{\omega}^\star - \tau_P K_P P(\theta^\star) \\ \quad\quad + \tau_P(\omega^d + (\alpha - I_N)\omega^{ref})\mathbf{1}_{N \times 1} = \mathbf{0}_{N \times 1} \end{cases} \tag{3.42}$$

The second equation in (3.42) implies that $\tilde{\omega}^\star = c\mathbf{1}_{N \times 1}$, where c is an arbitrary number. Then the first equation in (3.42) becomes $c(\beta W + \gamma \mathcal{L}_c \alpha)\mathbf{1}_{N \times 1} = \gamma \mathcal{L}_c \alpha(\hat{\omega}^\star - \omega^{ref}\mathbf{1}_{N \times 1})$. Since $\mathbf{1}_{N \times 1}$ does not lie in the range of $\mathcal{L}_c$, thus $c = 0$ and then $\tilde{\omega}^\star = \mathbf{0}_{N \times 1}$, which implies that $\omega^\star = \omega^{ref}\mathbf{1}_{N \times 1}$. This means that $\lim_{t \to \infty} \omega_i(t) = \omega^{ref}, \forall i \in N$. Furthermore, we have $\gamma \mathcal{L}_c \alpha(\hat{\omega}^\star - \omega^{ref}\mathbf{1}_{N \times 1}) = \gamma \mathcal{L}_c \alpha(\hat{\omega}^\star - \omega^\star) = \mathbf{0}_{N \times 1}$. Note that $U^{\omega^\star} = \alpha(\hat{\omega}^\star - \omega^\star)$, then $\gamma \mathcal{L}_c U^{\omega^\star} = \mathbf{0}_{N \times 1}$, which means that $\lim_{t \to \infty} (u_k^\omega - u_i^\omega) = 0, \forall i, k \in N$. This satisfies the input constraints (3.30) and also implies the accurate real power sharing (3.17). Referring to Theorem 3.2 in [57], (3.42) is solvable and has a unique solution $(\hat{\omega}^\star, \theta^\star, 0)$ if and only if $\max_{i,k} |\theta_i - \theta_k| < \frac{\pi}{2}, \forall i \in N, k \in \mathcal{N}_i$, which is typically satisfied in real applications as pointed out in [57]. Thus ii) and iii) are proved.

We now show (3.38) is locally exponentially stable. Similar to [57], by linearizing (3.38) around the equilibrium $(\hat{\omega}^\star, \delta^\star, \omega^{ref})$ we obtain

$$\begin{bmatrix} \Delta \dot{\hat{\omega}} \\ \Delta \dot{\delta} \\ \Delta \dot{\omega} \end{bmatrix} = A \begin{bmatrix} \Delta \hat{\omega} \\ \Delta \delta \\ \Delta \omega \end{bmatrix} \tag{3.43}$$

where A is given in (3.39), $\Delta\hat{\omega} = \hat{\omega} - \hat{\omega}^\star$, $\Delta\delta = \delta - \delta^\star$ and $\Delta\omega = \omega - \omega^\star$.

By calculation we obtain that $A\mathbf{e}_1 = 0$, where $\mathbf{e}_1 = \begin{bmatrix} \mathbf{0}_{1 \times N} & \mathbf{1}_{1 \times N} & \mathbf{0}_{1 \times N} \end{bmatrix}^T$ is an eigenvector of A corresponding to the eigenvalue 0, which indicates that A has at least one eigenvalue equal to 0. Next we will show that it has exactly one zero eigenvalue. Considering the linear state transformation (3.40), we obtain

$$\begin{bmatrix} \Delta\dot{\hat{\omega}} \\ \Delta\dot{\theta} \\ \Delta\dot{\omega} \end{bmatrix} = A' \begin{bmatrix} \Delta\hat{\omega} \\ \Delta\theta \\ \Delta\omega \end{bmatrix} \tag{3.44}$$

where

$$A' = \begin{bmatrix} -\gamma\mathcal{L}_c\alpha & \mathbf{0}_{N \times (N-1)} & \gamma\mathcal{L}_c\alpha + \beta W \\ \mathbf{0}_{(N-1) \times N} & \mathbf{0}_{(N-1) \times (N-1)} & \mathcal{R} \\ -\tau_P\alpha & -\tau_P K_p \mathcal{L}'_w & \tau_P(\alpha - I_N) \end{bmatrix}$$

with $\mathcal{L}'_w \in \mathbb{R}^{N \times (N-1)}$ being the weighted Laplacian matrix after state transformation, and $\Delta\theta = \theta - \theta^\star$.

Now we show that A' is a full-rank matrix. Consider

$$A' \begin{bmatrix} \Delta\hat{\omega} \\ \Delta\theta \\ \Delta\omega \end{bmatrix} = \mathbf{0}_{(3N-1) \times 1} \tag{3.45}$$

we have

$$\begin{cases} -\gamma\mathcal{L}_c\alpha\Delta\hat{\omega} + (\gamma\mathcal{L}_c\alpha + \beta W)\Delta\omega = \mathbf{0}_{N \times 1} \\ \mathcal{R}\Delta\omega = \mathbf{0}_{(N-1) \times 1} \\ -\tau_P\alpha\Delta\hat{\omega} - \tau_P K_p \mathcal{L}'_w\Delta\theta + \tau_P(\alpha - I_N)\Delta\omega = \mathbf{0}_{N \times 1} \end{cases} \tag{3.46}$$

The second equation in (3.46) implies that $\Delta\omega = k\mathbf{1}_{N \times 1}$, where k is an arbitrary number. Then the first equation in (3.46) becomes $\gamma\mathcal{L}_c\alpha\Delta\hat{\omega} = k(\gamma\mathcal{L}_c\alpha + \beta W)\mathbf{1}_{N \times 1}$. Since $\mathbf{1}_{N \times 1}$ does not lie in the range of $\mathcal{L}_c$, thus $k = 0$ and then $\Delta\hat{\omega} = \Delta\omega = \mathbf{0}_{N \times 1}$. Finally, consider the third equation in (3.46), we have $\tau_P K_p \mathcal{L}'_w\Delta\theta = \mathbf{0}_{N \times 1}$. As $rank(\mathcal{L}'_w) = N - 1$, which implies $\Delta\theta = \mathbf{0}_{(N-1) \times 1}$. As $\begin{bmatrix} \Delta\hat{\omega} & \Delta\theta & \Delta\omega \end{bmatrix}^T = \mathbf{0}_{(3N-1) \times 1}$ is the only solution of Eqn. (3.44), thus we conclude that A' is of full rank. Since the linear transform does not change the eigenvalues, hence A and A' have the same eigenvalues except that A has an extra zero eigenvalue. This indicates that A' is Hurwitz if and only if A has exactly one zero eigenvalue and all other eigenvalues are in the open left half complex plane. Thus the closed-loop system (3.38) is locally exponentially stable [148]. This completes the proof of Theorem 3.2. ■

3.4 Simulation Results

In order to test the designed secondary controller, a 220V (per phase RMS), 50 Hz islanded MG shown in Fig. 3.5 is considered as the test system. The simulation is

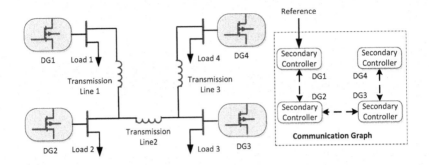

Figure 3.5: Simulation test system

Table 3.1: Parameters of primary controller and MG systems

		DG1		DG2		DG3		DG4
Model	τ_{P_1}	0.016	τ_{P_2}	0.016	τ_{P_3}	0.016	τ_{P_4}	0.016
	τ_{Q_1}	0.016	τ_{Q_2}	0.016	τ_{Q_3}	0.016	τ_{Q_4}	0.016
	k_{P_1}	6e-5	k_{P_2}	3e-5	k_{P_3}	2e-5	k_{P_4}	1.5e-5
	k_{Q_1}	4.2e-4	k_{Q_2}	4.2e-4	k_{Q_3}	4.2e-4	k_{Q_4}	4.2e-4
	k_{V_1}	1e-2	k_{V_2}	1e-2	k_{V_3}	1e-2	k_{V_4}	1e-2
Load	P_{1_1}	0.01	P_{1_2}	0.01	P_{1_3}	0.01	P_{1_4}	0.01
	P_{2_1}	1	P_{2_2}	2	P_{2_3}	3	P_{2_4}	4
	P_{3_1}	1e4	P_{3_2}	1e4	P_{3_3}	1e4	P_{3_4}	1e4
	Q_{1_1}	0.01	Q_{1_2}	0.01	Q_{1_3}	0.01	Q_{1_4}	0.01
	Q_{2_1}	1	Q_{2_2}	2	Q_{2_3}	3	Q_{2_4}	4
	Q_{3_1}	1e4	Q_{3_2}	1e4	Q_{3_3}	1e4	Q_{3_4}	1e4
Line	$B_{12}=10\Omega^{-1}$, $B_{23}=10.67\Omega^{-1}$, $B_{34}=9.82\Omega^{-1}$							

conducted in the MATLAB Simulink environment. This islanded MG consists of four DGs, four respective local loads, and three transmission lines. The parameters of the whole test system are summarized in Table 3.1 and Table 3.2. To facilitate the illustration, in this example we choose the same communication graph as the physical connection graph. The communication graph for the distributed secondary controllers is also shown in Fig. 3.5. The reference voltage and frequency values are only known to DG1, i.e., $g_1{}^V = g_1{}^\omega = 1$, $g_k{}^V = g_k{}^\omega = 0, k = 2, 3, 4$. All the parameters satisfy the conditions in Theorems 3.1 and 3.2.

This simulation can be divided into 4 stages:

Stage 1 (0 - 5s): Only primary control is activated

Stage 2 (5 - 10s): Secondary control is activated

Stage 3 (10 - 30s): Constant load $L_c = 1 \times 10^4 + j1 \times 10^4 W$ is added to Load 4

Stage 4 (30 - 40s): Load L_c is removed from Load 4

The simulation results are shown in Figs. 3.6-3.9. As seen from Fig. 3.6 and 3.8,

Table 3.2: Parameters of secondary controller

	DG1		DG2		DG3		DG4	
Frequency Controller	α_1	0.2	α_2	0.2	α_3	0.2	α_4	0.2
	β_1	250	β_2	250	β_3	250	β_4	250
	γ_1	500	γ_2	500	γ_3	500	γ_4	500
Voltage Controller	k_1	100	k_1	100	k_1	100	k_1	100
	k_2	10	k_2	10	k_2	10	k_2	10
	α_1	$\frac{1}{3}$	α_1	$\frac{1}{3}$	α_1	$\frac{1}{3}$	α_1	$\frac{1}{3}$
	α_2	$\frac{1}{2}$	α_2	$\frac{1}{2}$	α_2	$\frac{1}{2}$	α_2	$\frac{1}{2}$
Reference	$V^{ref} = 310V,\ \omega^{ref} = 50\text{Hz}$							

during stage 1, due to the droop function in the primary control, the voltage amplitudes of 4 DGs fall down to different values while the frequency can synchronize to a common value (49.72Hz). Unfortunately, both voltage and frequency deviate from their reference values, hence they need to be restored in the secondary control layer. When our distributed secondary control is activated at $t = 5s$, both voltage and frequency can quickly restore to their reference values respectively (V^{ref}=310V, ω^{ref}=50Hz). The steady-state frequencies of the four DGs remain at 50Hz no matter new constant load L_c is connected to or disconnected from DG4, even though there are transient deviations. This result shows that the designed distributed secondary controller can eliminate the voltage and frequency deviation caused by the primary control.

The real power outputs of these tested four DGs are shown in Fig. 3.7. Before the secondary control is activated (stage 1), the real power sharing is well achieved by the primary control, i.e., $P_1 : P_2 : P_3 : P_4 = \frac{1}{k_{P_1}} : \frac{1}{k_{P_2}} : \frac{1}{k_{P_3}} : \frac{1}{k_{P_4}} = 1 : 2 : 3 : 4$. When the secondary control is activated (from 5s), real power is still well shared according to the designed droop gains regardless of load increasing at stage 3 or decreasing at stage 4. The secondary frequency control inputs are shown in Fig. 3.9. Clearly these simulation results validate that secondary frequency control inputs are equal to each other in the steady state, i.e., $(u_1{}^\omega)_s = (u_2{}^\omega)_s = (u_3{}^\omega)_s = (u_4{}^\omega)_s$.

As mentioned in the introduction, there are limited approaches to solve voltage and frequency restoration problems in a distributed way. As far as we know, relevant results are only available in [54]-[59]. Among these approaches, only [54] and [55] consider similar problems to that of this paper. However in [54], frequency restoration is not considered. So here we just make comparison with the method proposed in [54] in voltage restoration by also applying it to the above MG system. The control gains are set as $k_1 = 100, k_2 = 40$ in both our proposed method and the controller in [54]. The simulation results of stage 2 for the period [4.8s, 7.5s] are shown in Fig. 3.10. It is clear from Fig. 3.10 that our proposed method has settling time of about 1.5 seconds, while the method in [54] requires 2.5 seconds for the responses to settle down, which is about 166% of our settling time. This actually makes sense as our scheme ensures the convergence within finite time.

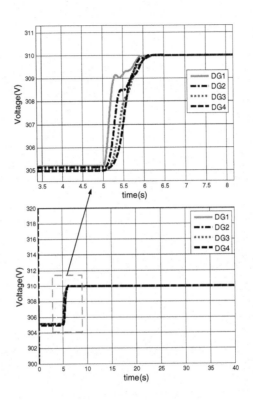

Figure 3.6: The voltage output of the test islanded MG

Acknowledgments

©2017 IEEE. Required, with permission, from Fanghong Guo, Changyun Wen, Jianfeng Mao, and Yong-Duan Song, "Distributed Secondary Voltage and Frequency Restoration Control of Droop-Controlled Inverter-Based Microgrids," *IEEE Transactions on Industrial Electronics*, vol. 62, no. 7, pp. 4355 - 4364, 2015.

©2017 IEEE. Required, with permission, from Fanghong Guo, Changyun Wen, "Distributed Control Subject to Constraints on Control Inputs: A Case Study on Secondary Control of Droop-Controlled Inverter-Based Microgrids," 2014 IEEE 9th Conference on Industrial Electronics and Applications (ICIEA), pp. 1119 - 1124, Hangzhou, China, 2014.

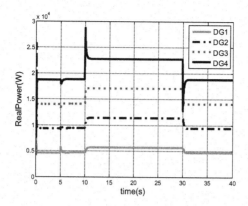

Figure 3.7: The real power output of the test islanded MG

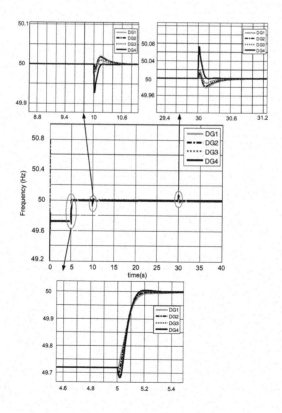

Figure 3.8: The frequency output of the test islanded MG

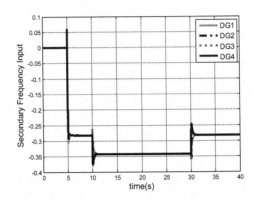

Figure 3.9: The secondary frequency control input of the test islanded MG

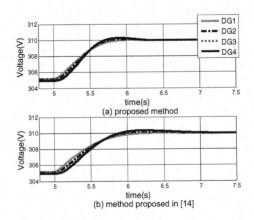

Figure 3.10: Comparison between the proposed method and the approach in [54]

Chapter 4

Distributed Voltage Unbalance Compensation

In the last chapter, distributed voltage and frequency restoration control is addressed. In this chapter, we continue to employ the distributed control strategy in the secondary control layer but to another research topic, i.e., voltage unbalance compensation.

Conventionally, such a problem is usually solved in a centralized way. In this chapter, we try to resolve it in a distributed manner. By letting each DG share the compensation effort cooperatively, unbalanced voltage in sensitive load bus (SLB) can be compensated. The concept of contribution level for compensation is first proposed for each local DG to indicate its compensation ability. A two-layer secondary compensation architecture consisting of a communication layer and compensation layer is designed for each local DG. A totally distributed strategy involving information sharing and exchange is proposed, which is based on finite-time average consensus and newly developed graph discovery algorithm. This strategy does not require the whole system structure as a prior and can detect the structure automatically. The proposed scheme not only achieves similar voltage unbalance compensation performance to the centralized one, but also brings some advantages, such as communication fault tolerance and plug-and-play property. Case studies including communication failure, contribution level variation, and DG plug-and-play are discussed and tested to validate the proposed method.

4.1   Introduction

Voltage unbalance is considered one of the power quality problems, which is mainly caused by the unbalanced loads, incomplete transposition of transmission line, and open delta transformer connections and so on [149]. By definition, any difference that exists in the three voltage magnitudes and/or a shift in the phase separation from 120 degrees is said to have unbalanced voltage [150]. The unbalanced voltage output can be harmful to its connected loads, especially for induction motors.

A distributed cooperative secondary control architecture for voltage unbalance compensation is proposed in this chapter. In this architecture, we decompose the centralized secondary controller into distributed ones. Each controller is located at a local DG unit with an architecture of two layers, *namely*, communication layer and secondary compensation layer. By communicating with its neighboring controllers, each local controller can share the compensation efforts cooperatively to compensate for the unbalanced voltage in SLB. To consider the compensation ability of each DG, contribution level is assigned based on its operational conditions. With this, it is not necessary that all the DGs need to participate in the compensation.

In our approach, we first propose a totally distributed finite-time average consensus algorithm, which does not need to know the whole system structure as a prior and is able to detect the structure by each agent automatically. Then we employ the algorithm for the voltage unbalance compensation, which can discover and share the global information within finite steps. Furthermore, considering possible communication failure of some local controllers as well as plug-and-play of certain DGs, a distributed cooperative secondary control scheme is proposed. Several case studies are conducted to validate our proposed method. It is further illustrated that system stability and acceptable performance are ensured as long as the consensus time is within certain bounds, which also shows that finite-time consensus is necessary for distributed VUC.

4.2   Distributed Cooperative Secondary Control Scheme for Voltage Unbalance Compensation

In this section, a distributed cooperative secondary control scheme for voltage unbalance compensation (VUC) will be presented. For completeness and comparison, the centralized VUC approach is briefly introduced first.

4.2.1   A centralized VUC approach

One key factor called voltage unbalance factor (VUF) is usually employed to describe the unbalanced voltage, which is defined as $VUF = V_2/V_1$, where V_1 and V_2 are the voltage magnitudes of the positive and negative sequence, respectively [64]. The higher the VUF, the more the unbalanced voltage output.

In this chapter, the VUC in the SLB is considered. One typical solution to this problem can be found in [64] and [65]. Its main idea can be summarized as follows: the SLB voltage V_{abc} is sampled and transformed to the dq frame through the abc/dq transform. By applying the symmetrical decomposition and the utilization of two second-order low pass filters, the positive and negative sequence components are extracted, which are used to calculate the VUF. Then the error between the calculated VUF and reference VUF is fed to a PI controller. Afterwards, the unbalance compensation reference (UCR) UCR_{dq} is generated by multiplying the PI controller output with the negative sequence component. Lastly the UCR is shared by the DGs in the MG equally, i.e.,

$$UCR_{dq_i} = \frac{1}{N}UCR_{dq}, \ i = 1, \cdots, N \qquad (4.1)$$

where UCR_{dq_i}, N are the compensation reference of the ith DG and the total number of DGs in the MG, respectively.

Note that the above compensation effort is shared equally by all the DGs and it is realized in a centralized controller. The communication burden and cost of the centralized controller may increase greatly when the number of the DGs becomes larger especially when the DGs are distributed sparsely. In addition, if this centralized controller fails to work, the compensation may not be achieved. Also, when the MG structure changes, this centralized controller may need to be redesigned.

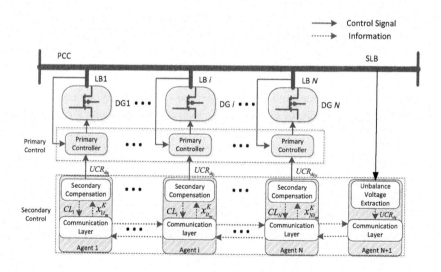

Figure 4.1: Distributed voltage unbalance compensation in an islanded MG

4.2.2 Distributed VUC approach

In this subsection, we present an FACA-based distributed secondary control scheme for VUC in an islanded MG system. The block diagram of the proposed control structure is shown in Fig. 4.1. In our design, we first treat the local secondary controller as an "agent," and each agent is assigned a unique ID. The voltage in SLB is monitored and extracted by a particular agent (e.g., agent $N+1$ is assigned in Fig. 4.1). Then we divide the agent into two layers, i.e., communication layer and compensation layer. The compensation layer sends a secondary compensation reference signal UCR_{dq_i} (illustrated with solid lines) to the primary control and returns its contribution level CL_i (which will be introduced below) to the communication layer, while the communication layer is mainly responsible for exchanging information with neighbors to obtain the global information cooperatively and then sends it to the compensation layer.

4.2.2.1 Preliminary setup

In contrast to the average UCR sharing strategy proposed for the centralized approach in [65], we allow each DG to have different levels of contributions in compensating for unbalanced voltage, depending on its operating conditions. To do this, we first assign a contribution level (CL) for each DG as one of the following levels: zero contribution, small contribution, medium and high contributions represented by numerical numbers $0, 1, 2, 3$, respectively. For example, during a certain period, if the ith DG is short of power or operates abnormally, then it is not able to participate in the secondary compensation, then its CL can be assigned as $CL_i = 0$; if the loads connected to local bus of the ith DG are not critical to having unbalanced voltage, which means that the ith DG can share more unbalanced compensation effort; then its CL can be assigned as $CL_i = 3$, otherwise $CL_i = 1$ or $CL_i = 2$. Note that the CL of a DG can change according to its local operating situations. With this assignment, each DG compensates for the unbalanced voltage in the SLB cooperatively.

Besides, in this chapter, the communication fault (CF) in certain DGs is also considered. To facilitate the illustration, the communication fault considered occurs at the communication node, i.e., controller fault or failure, rather than in the communication line. Such a fault is one type of communication fault based on the definition in [151]. If the neighboring agents receive no communication response from agent i, then we claim a CF occurs in agent i, and set $CF_i = 1$, otherwise $CF_i = 0$.

We also allow new agents to be added and existing agents to be removed at any time. For example, some DGs are newly installed in the MG system; some old DGs are temporarily or permanently uninstalled from the MG. This requires that the scheme be adaptive to dynamic MG structure.

4.2.2.2 Distributed VUC

The communicated information among the $N+1$ agents in the secondary communication layer include the unbalance compensation reference (UCR_{dq}) for the SLB and the contribution level ($CL_i, i = 1, \cdots, N$). Different from the centralized "one-to-all" communication structure, in the distributed scheme, the communication is among the agents [24]-[25]. Each agent only has access to the local information, instead of the entire global information. That is, initially UCR_{dq} and CL_i are only known by agents $N+1$ and $i, i = 1, \cdots, N$, respectively. Besides, each agent can only communicate with its immediate neighboring agents, which can be chosen as the same as the physically connected neighbors.

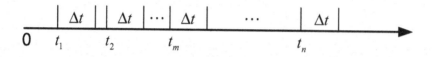

Figure 4.2: Illustration of secondary compensation in time domain

The secondary compensation in time domain is illustrated in Fig. 4.2. Each agent samples its local information and sets them as the communication initial value at t_m, $m = 1, 2, \cdots$ (which will be presented in detail later). Based on the FACA in (2.2), it is ensured that information exchange can be finished within finite-time steps and each communication period is a constant Δt in a given communication graph. Once the average consensus is reached, each agent starts the next round of communication.

Let $x^l_{it_m}$ be the communication information state of agent i at iteration l during the communication period $[t_m, t_m + \Delta t]$. Its initial value $x^0_{it_m}$ is set as

$$x^0_{it_m} = \begin{bmatrix} a_i UCR_{dq}(t_m) & CL_i(t_m) \end{bmatrix}^T \tag{4.2}$$

where $UCR_{dq}(t_m)$, $CL_i(t_m)$ are the sampled local information by agent $N+1$ and agent i at t_m, respectively, a_i indicates whether agent i is assigned to monitor unbalance voltage information from SLB. If assigned, then $a_i = 1$, otherwise $a_i = 0$. In Fig. 4.1, agent $N+1$ is assigned, thus $a_{N+1} = 1$, $a_i = 0, i = 1, \cdots, N$. Also we define $CL_{N+1} = 0$.

According to Lemma 2.1, after finite K steps, the communicated information of each agent $\forall i, j = 1, \cdots, N$ reaches consensus as

$$x^K_{it_m} = x^K_{jt_m} = \begin{bmatrix} \dfrac{a_{N+1} UCR_{dq}(t_m)}{N+1} & \dfrac{\sum_{k=1}^{N+1} CL_k(t_m)}{N+1} \end{bmatrix}^T \tag{4.3}$$

Then the communication layer sends the consensus information to the secondary compensation layer. Motivated by (4.1), the distributed secondary compensator for the voltage compensation is designed as follows:

$$UCR_{dq_i} = \frac{CL_i}{\hat{S}_i} \widehat{UCR_{dq_i}} \tag{4.4}$$

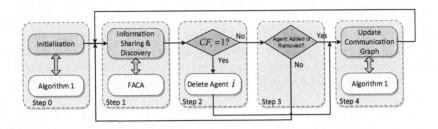

Figure 4.3: Flowchart of distributed cooperative secondary control scheme

where $\hat{S}_i = \{x_{it_m}^K\}_2$, $\widehat{UCR_{dq_i}} = \{x_{it_m}^K\}_1$ denoting consensus information obtained by agent i. Here $\{y\}_k$ denotes the kth element of vector y.

Note that once the average consensus is reached, each agent updates UCR_{dq_i} according to Eqn. (4.4) and starts the next round of communication. During the period $[t_m, t_m + \Delta t]$, UCR_{dq_i} is held to be the previously updated value. Then the obtained unbalance compensation reference UCR_{dq_i} is sent to the primary control layer to realize the voltage unbalance compensation. The detailed information about the primary control layer can be found in [64] and [65].

Now we summarize our distributed cooperative secondary control scheme design.

4.2.2.3 Distributed cooperative secondary control scheme (DCSCS) design

The flowchart of our proposed distributed cooperative secondary control scheme is shown in Fig. 4.3, with each corresponding step described as follows:

Step 0: Initialization and graph discovery: As a starting point, the communication graph of all agents is pre-designed as connected.[1] Using Algorithm 1, each agent can get the information of the whole communication graph including the number of agents $N + 1$ and the Laplacian matrix $\mathcal{L}$, in less than N steps. Then certain available numerical methods can be used to calculate the nonzero eigenvalues of $\mathcal{L}$.

Step 1: Information sharing and compensation: At this step, each agent first uses the obtained eigenvalues to calculate the update gains according to (2.2) and then apply FACA protocol (2.1) for information sharing and discovery. Then each agent calculates the compensation reference according to (4.4) and sends it to its primary control layer.

Step 2: CF monitoring: At this step, agent $j, j = 1, \cdots, N + 1$ needs to check

---

[1]This is easily implementable, as each agent can choose the same communication graph as their physical connection graph at the initial step.

whether any CF occurs between itself and its neighbors $i, j \in \mathcal{N}_i$ in the communication layer. If yes, agent j should delete i from its neighbor table and then go to Step 4. If no, go to Step 3.

Step 3: Plug-in and plug-out reconfiguration: At this step, each agent needs to check whether there is any agent added to or removed from the grid. If yes, execute the *Graph Reconfiguration* rule described below, and then go to Step 4 to update the communication graph; otherwise go to Step 1.

Step 4: Graph updating: At this step, all agents need to update the communication graph using Algorithm 2.1 and then go to Step 1.

Graph reconfiguration: If an agent (agent n) is newly added in, it tries to find its nearest neighbors, gets permission from them and then adds them in its neighbor list. If an agent (say agent i) is removed, its neighboring agent j, $j \in \mathcal{N}_i$ will delete agent i from its neighborhood list $\mathcal{N}_j$ and also tries to set up communication with other agent k, $k \in \mathcal{N}_i \backslash j$, which is also the neighbor of agent i. If $k \in \emptyset$, i.e., no other neighbor of agent i exists, then nothing is needed to be done but just deleting agent i.

Remark 4.1 *With the use of limited communication among the neighboring DGs, we can still achieve similar compensation performance to those in [61]-[63]. Compared to existing results [61]-[65], [152], this scheme has the following advantages: 1) The proposed DCSCS is totally distributed, with no preliminary knowledge of the system such as the number of agents which is assumed to be known in [152]; 2) The communication fault can be detected by each agent individually, which improves the reliability of the whole system; 3) It also brings some other advantages such as plug-and-play property, i.e., it allows new agents to be added and existing agents be to removed.*

4.2.2.4 Stability analysis of distributed VUC

In our designed distributed communication protocol, the distributed FACA is employed. Compared to other distributed protocols in [153], [152], [154]-[156], the average consensus in our scheme can be reached in a finite-time Δt, which is much shorter than that in conventional average consensus algorithm in [152]. In practice, the distributed communication can be applied to wireless networks, such as ZigBee, WiFi, and cellular communication networks [152]. For the long-range low-delay network such as cellular communication network, the communication time delay of each iteration, ΔT, is usually negligible as pointed out in [152]. However, note that the convergence time Δt can be simply estimated as $\Delta t = K \times \Delta T$, where K is determined by the communication graph. When the number of DGs becomes larger, or there exists larger communication delay between each agent, the convergence time Δt becomes non-negligible. During the period $[t_m, t_m + \Delta t]$, UCR_{dq_i} is held to be previously updated value according to Eqn. (4.4). Such a Δt then has an impact on the stability and performance of secondary VUC. Following [63] and [65], we will study the stability and performance of the whole system in the case studies of the next section by considering different consensus time Δt. It is illustrated that system stability and acceptable performance are ensured as long as the consensus time Δt is within certain bounds. Such studies also show that finite-time consensus is necessary for distributed VUC.

4.3 Case Studies

In order to validate the proposed distributed control scheme, a simulation test model is built in the Matlab® Simulink environment. The islanded MG system with unbalanced load in the SLB is considered as the test system, which is shown in Fig. 4.4. This MG system consists of three regular DGs (DG1, DG2 and DG3) and one backup DG (DG4) with different pre-assigned contribution levels ($CL_1 = 1, CL_2 = 2, CL_3 = 3, CL_4 = 1$) and different power ratings ($S_1 : S_2 : S_3 : S_4 = \frac{1}{m_{P_1}} : \frac{1}{m_{P_2}} : \frac{1}{m_{P_3}} : \frac{1}{m_{P_4}} = 1 : 2 : 3 : 4$), where each $m_{P_i}, i = 1, \cdots, 4$ is the frequency droop gain and usually chosen as the inverse proportion of the power rating. A balanced load (Load 1 Z_B) and an unbalanced load (Load 2 Z_{UB}) are connected to the SLB. The primary control layer design is adopted from [65] and its key design parameters are listed in Table 4.1. The parameters of the secondary controller as well as the MG system are summarized in Table 4.2. Note that the reference VUF is set as $VUF^{ref} - 0.5\%$ here by considering the standard *ANSI C84.1-1995* [157] and the measurement errors existing in practical industrial environments.

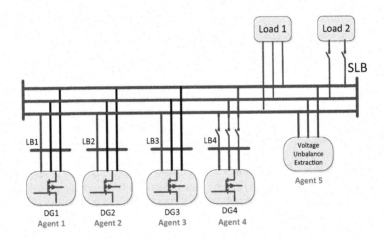

Figure 4.4: Simulation test system

Table 4.1: Parameters of DG and its primary controller

		DG1 / DG2		DG3 / DG4
DG	V_{DC}	700V	V_{DC}	700V
	f_s	10kHz	f_s	10kHz
Primary Controller	K_{pv}	2	K_{pv}	2
	K_{rv}	0.5	K_{rv}	0.5
	ω_{cv}	10	ω_{cv}	10
	K_{pi}	10	K_{pi}	10
	K_{ri}	10	K_{ri}	10
	m_P	6e-4 / 3e-4	m_P	2e-4 / 1.5e-4
	n_Q	1.3e-3	n_Q	1.3e-3
	R_v	2 / 1	R_v	$\frac{2}{3}$ / $\frac{1}{2}$
	L_v	8e-3 / 4e-3	L_v	$\frac{8}{3}$e-3 / 2e-3

Table 4.2: Parameters of secondary controller and MG systems

		DG1 / DG2	DG3 / DG4
Controller		$K_p = 1, K_i = 20, VUF^{ref} = 0.5\%$	
LCL Filter	L_f	1.5e-3H	1.5e-3H
	C_f	50μF	50μF
	R_f	0.05Ω	0.05Ω
Transmission Line	R_L	0.23Ω / 0.35Ω	0.2Ω / 0.2Ω
	X_L	3.18e-4H / 1.87e-3H	2.31e-3H / 2.31e-3H
Load 1		Z_B=50+j12.57Ω (per phase)	
Load 2		Z_{UB}=30Ω	

4.3.1 Testing of the overall distributed control system under various cases

The overall system consists of the DGs, loads, transmission lines as well as primary controllers and proposed secondary controllers. In order to consider various cases, the whole simulation is divided into 9 stages:

Stage 1 (0-2s): System operates in a balanced steady state during which Load 1 is connected to the SLB

Stage 2 (2-5s): Load 2 is connected to the SLB

Stage 3 (5-7s): Communication fault occurs in DG2

Stage 4 (7-9s): Communication fault is cleared in DG2

Stage 5 (9-11s): Contribution level changes in DG3

Stage 6 (11-13s): Contribution level changes in DG1

Stage 7 (13-15s): DG4 is plugged in

Stage 8 (15-17s): DG4 is removed

Stage 9 (17-18s): Load 2 is disconnected from the SLB

Our proposed scheme is applied to the system experiencing the above stages. After Step 0 (Initialization) in Fig. 4.3, each agent obtains the graph information such as the Laplacian matrix $\mathcal{L}$ as

$$\mathcal{L} = \begin{bmatrix} 1 & -1 & 0 & 0 \\ -1 & 2 & -1 & 0 \\ 0 & -1 & 2 & -1 \\ 0 & 0 & -1 & 1 \end{bmatrix}$$

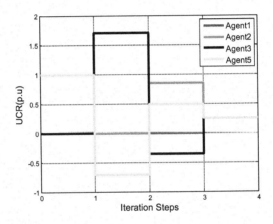

Figure 4.5: The average consensus process of UCR

Then each agent calculates the nonzero eigenvalues of $\mathcal{L}$ as $\lambda_2 = 0.5858$, $\lambda_3 = 2$, $\lambda_4 = 3.4142$, which also indicates that only 3 steps are needed for each agent to reach consensus. The corresponding updating gains of FACA for agent i can be immediately determined according to (2.2). Taking the first element of the state x_{it_m} as an example, it actually indicates the unbalance compensation reference of the SLB UCR_{dq}, which is monitored and extracted by agent 5 in this test system.

To illustrate this process clearly, the iteration result of this state is shown in Fig. 4.5. As seen from this figure, the initial values of this state for the 4 agents are 0, 0, 0, 1 p.u., respectively. After 3 steps of communication, the state values of the 4 agents reach to a consensus value 1/4 p.u., which validates the effectiveness of the applied finite time average consensus algorithm.

After each time consensus, agent i calculates the ith compensation effort according to (4.4), and then sends it to the primary control layer.

It is worthy to point out that as pointed out in [152], the communication time delay is usually negligible within a long-range low-delay network such as a cellular communication network. Thus, $\Delta T \approx 0$, which leads to $\Delta t \approx 0$. Because of high computational speed in the simulation example, an ideal communication case can be treated and marked as $\Delta t = 0$. In this section, such an ideal situation $\Delta t = 0$ is considered in all the case studies except for the case of studying system stability and performance.

The simulation results for the 9 stages are shown in Figs. 4.6-4.9. As seen from stages 2-9 in Fig. 4.6, the voltage in the SLB is guaranteed to be balanced with the VUF being less than 1% after the compensation at the sacrifice of the unbalanced voltage output in DG1, DG2, and DG3.

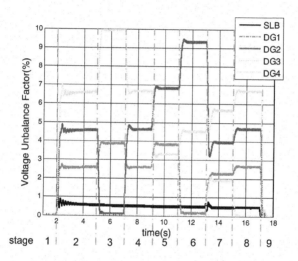

Figure 4.6: Voltage unbalance factors of each DG and SLB

The real and reactive power outputs of the MG system are shown in Fig. 4.7. The frequency outputs of the MG system are shown in Fig. 4.8. During stage 1, the frequency outputs of DG1, 2 and 3 are synchronized to 49.9 Hz when Load 1 is connected. When Load 2 is connected, the frequency drops to 49.7 Hz due to the droop function property. Also observed from Fig. 4.8, the frequency increases a bit due to the plug-in of DG4. It is observed that the steady-state frequency deviates from the nominal 50 Hz. To maintain at 50 Hz, one possible solution is to apply the approach of frequency restoration such as the method presented in Chapter 3, which is another topic and can be addressed separately. The unbalanced compensation references UCR_{d_i} and UCR_{q_i} produced by the secondary compensation controller are also shown in Fig. 4.9.

The amplitudes of negative sequence current of each bus are demonstrated, as shown in Fig. 4.10. It is observed that during stage 2-8, the negative sequence current in SLB is shared by each DG in different proportions according to the pre-designed compensation effort. Such a phenomenon can be explained as follows. As the voltage amplitude at SLB can be treated as constant, then constant unbalanced load can result in constant negative sequence current. If this current flows through SLB, it will result in unbalanced voltage output. Under our compensation principle, this negative sequence current can flow to the local buses of participating DGs. According to the well-known Kirchhoff's Current Law, the sum of negative sequence current in each local bus is equal to that in SLB.

In the following, 3 different case studies will be discussed.

4.3.1.1 Case A: Communication failure

During stage 3, a communication fault occurs in DG2 at $t = 5s$. By applying the designed DCSCS, each agent can autonomously reconfigure a communication graph, which is shown in Table 4.3. In the graph DG2 is excluded. Also observed from Fig. 4.9, its UCR is $UCR_2 = 0$, which means only DG1 and DG3 participate in the compensation. The updating gains of FACA are updated according to the new graph. Note that each agent only needs 2 steps to reach consensus when exchanging the information. From the simulation results of stage 3 in Fig. 4.6, we can see that although a CF occurs in DG2, the VUC in SLB is still achieved by our proposed DCSCS. It is also noted that because of the absence of DG2, the VUFs of DG1 and DG3 increase to almost 10% and 4%, respectively.

At stage 4, the communication fault is cleared from DG2. As seen from Figs. 4.6-4.9, the whole system operates normally, the same as that at stage 2.

Table 4.3: Communication graph reconfiguration in Case A

	Graph	Laplacian Matrix
Before		$\mathcal{L} = \begin{bmatrix} 1 & -1 & 0 & 0 \\ -1 & 2 & -1 & 0 \\ 0 & -1 & 2 & -1 \\ 0 & 0 & -1 & 1 \end{bmatrix}$
After		$\mathcal{L} = \begin{bmatrix} 1 & -1 & 0 \\ -1 & 2 & -1 \\ 0 & -1 & 1 \end{bmatrix}$

4.3.1.2 Case B: Contribution level variation

During stages 5 and 6, the contribution levels of certain DGs vary during some time periods. From $t = 9s$ to $t = 11s$, DG3 changes its contribution level to $CL_3 = 1$, which means its local bus may have sensitive load connected and cannot share more effort in the SLB compensation. During this period, the VUF of DG3 decreases from 6% to 3% while those of DG1 and DG2 increase to 4% and 7%, respectively. From $t = 11s$, DG1 does not participate in the compensation and sets its CL as $CL_1 = 0$. Therefore during this period, only DG2 and DG3 contribute the compensation effort cooperatively. At this stage, DG1 almost has the balanced voltage output at its local bus while DG2 and DG3 have more unbalanced voltage outputs with VUFs being 9% and 4.5%, respectively. The simulation results at stages 5 and 6 of Fig. 4.6 show that each DG can share the compensation effort dynamically with different contribution levels.

4.3.1.3 Case C: Backup DG plug-and-play

During stages 7 and 8, the plug-and-play property of the proposed method is tested. The backup DG (DG4) is assumed to be connected to the PCC from $t = 13s$ to $t = 15s$ with a contribution level $CL_4 = 1$. The communication topology can be reconfigured by each agent autonomously according to the *Graph Reconfiguration*, which is shown in Table 4.4. Similar to Case A, the updating gains of FACA are updated according to the new graph and each agent needs 4 steps to reach consensus in FACA algorithm (as the new Laplacian matrix $\mathcal{L}$ has 4 distinct nonzero eigenvalues). During the period $[13s, 15s]$, the VUFs of DG1, DG2, and DG3 decrease to 2%, 4%, and 5.5%, respectively, at the expense of unbalanced output of DG4. DG4 is disconnected from the islanded MG system from $t = 15s$. As can be seen from Fig. 4.6, the VUFs of DG1, DG2, and DG3 return to the values during stage 4. Note that stage 8 is equivalent to the case that DG4 is unavailable, which can be handled by our proposed scheme.

Also observed from Fig. 4.8, the frequency increases a bit due to the plug-in of DG4. However, there exists a transient response in frequency (also in real and reactive power outputs in Fig. 4.7) when DG4 is connected. The reason can be interpreted as follows. As the frequency and the voltage angle of DG4 are different from those of the islanded MG system when it is suddenly connected, there will be a transient process before reaching synchronization. These transient power fluctuations may be reduced by the well-known grid synchronization phase-locked loop (PLL) method before DG4 is connected [158]. Although it is out of the scope of this chapter, we feel that such an issue is an interesting topic worthy of consideration as a future work.

4.3.2 System stability and performance

In this subsection, the stability and performance of the overall nonlinear system is investigated through simulation studies by considering different consensus time, *namely* 0ms, 1ms, 5ms, and 7ms. The zero-order hold (ZOH) block is used to implement the holding of UCR_{dq_i} over period $[t_m, t_m + \Delta t]$ with consensus time Δt as the

Table 4.4: Communication graph reconfiguration in Case C

	Graph	Laplacian Matrix
Before		$\mathcal{L} = \begin{bmatrix} 1 & -1 & 0 & 0 \\ -1 & 2 & -1 & 0 \\ 0 & -1 & 2 & -1 \\ 0 & 0 & -1 & 1 \end{bmatrix}$
After		$\mathcal{L} = \begin{bmatrix} 1 & -1 & 0 & 0 & 0 \\ -1 & 2 & -1 & 0 & 0 \\ 0 & -1 & 3 & -1 & -1 \\ 0 & 0 & -1 & 1 & 0 \\ 0 & 0 & -1 & 0 & 1 \end{bmatrix}$

sampling period of the ZOH block. The VUF outputs of all the DGs and SLB under various Δt are shown in the figures given in Table 4.5. Obviously the communication consensus time affects the voltage unbalance compensation. It is observed from the simulation results that the system remains stable for $\Delta t \leq 5$ms. However, it becomes unstable when $\Delta t \geq 7$ms. As pointed out in [152], the communication time delay is usually negligible for cellular communication networks. Thus the convergence time is very small, compared to the stability margin of 7ms. So this system is surely in stable operation.

From these results, it can be concluded that our proposed scheme ensures system stability as long as the consensus time is within certain bound that can be determined when the scheme is applied to a particular system. These results also show that finite-time consensus is necessary for distributed VUC in order to have Δt meet the bound.

It is also noted that the performance of VUC deteriorates when the consensus time becomes larger and larger. The case that $\Delta t = 0$ is an ideal case which is used as a basis for comparisons. By comparing the case that $\Delta t = 1ms$ to such an ideal case, little difference is observed in their performances. Thus we can conclude that acceptable compensation performances can be achieved with consensus time Δt less than $1ms$ for this application example. Such studies and the observed results provide us design guidelines on the choice of Δt in designing distributed VUC.

4.3.3 Comparisons with centralized secondary control in [65]

For comparison, the centralized secondary UCR sharing strategy (4.1) proposed in [65] is also applied in our simulation test system. The simulation results are shown in Fig. 4.11. We now compare Fig. 4.11 with Fig. 4.6 in the time period from 2s to 5s. Clearly the two responses of VUFs in the SLB almost have the same transient and steady-state performances. In addition, from Fig. 4.7, we also conclude that our pro-

Table 4.5: Voltage unbalance factors output with different consensus time

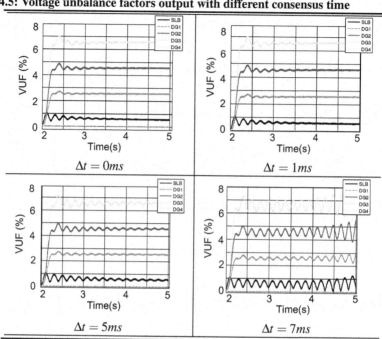

$\Delta t = 0ms$ $\Delta t = 1ms$

$\Delta t = 5ms$ $\Delta t = 7ms$

posed distributed secondary VUC almost has no impact on power sharing even with the plug-and-play of certain DG in Case C. Thus it has achieved a similar performance to that in [65]. These results illustrate that our proposed distributed secondary compensation strategy performs as well as the centralized controller in the compensation. On the other hand, it can dynamically share the compensation effort among the DGs and have the property of plug-and-play, which cannot be realized in the conventional centralized controller.

We now consider the failure situation of the secondary communication layer. In this case study, as mentioned before, the communication fault is supposed to happen in the communication node, i.e., controller fault. As shown in Case A above, our proposed distributed control strategy guarantees balanced voltage output in the SLB if some secondary communication layers fail to work. Unfortunately, when communication failure occurs in the centralized compensation sharing controller, Fig. 4.11 shows that it is unable to compensate for the unbalanced voltage from 5s to 9s.

4.3.4 *Distributed voltage unbalance compensation using negative sequence current feedback*

Note that in (4.4), the compensation is realized in an open-loop fashion. In this subsection, the negative sequence current feedback is utilized to realize the unbalance compensation reference sharing in a closed-loop way. A distributed proportion integral (PI) negative sequence current controller is designed in each local DG to let them share the negative sequence current according to their own power rating and compensation ability.

The proposed compensation method mainly consists of two parts, i.e., SLB voltage unbalance compensation and negative sequence current feedback, as shown in Fig. 4.12.

As illustrated in Fig. 4.12, the negative sequence current components I_{id}^-, and I_{iq}^- at each local bus are extracted by using a similar symmetrical decomposition method. Afterwards, motivated by the distributed PI controller design idea in Chapter 3, a distributed PI negative sequence current sharing controller is designed in each local secondary compensation layer as follows,

$$UCR_i^I = K_{Pi}e_i^{Neg} + K_{Ii} \int e_i^{Neg} \tag{4.5}$$

$$e_i^{Neg} = \sum_{j \in \mathcal{N}_i} (w_i I_i^{Neg} - w_j I_j^{Neg}) \tag{4.6}$$

where K_{Pi}, K_{Ii} are the proportional and integral gains, respectively, w_i is the negative current sharing gain, which can be chosen according to the DG power rating inverse ratios, i.e., $\frac{w_i}{w_j} = \frac{P_j^{rating}}{P_i^{rating}}$, I_i^{Neg} is the amplitude of the negative sequence current of the ith DG, satisfying $I_i^{Neg} = \sqrt{I_{id}^{-2} + I_{iq}^{-2}}$, $\mathcal{N}_i$ is the neighboring set of the ith DG.

Finally, summing up the generated two compensation signals UCR^{SLB}, UCR_i^I, and multiplying the sum signal with the negative sequence voltage at SLB V_d^-, V_q^- produces the unbalance compensation reference signal UCR_{id}, UCR_{iq}, respectively. The obtained reference signal is sent to the primary control layer to realize the VUC.

The whole simulation can be divided into 5 stages:

Stage 1 (0-1s): Black start stage during which Load 1 is connected to the SLB

Stage 2 (1-2s): System operates in a balanced steady state

Stage 3 (2-8s): Load 2 is connected to the SLB

Stage 4 (8-15s): Distributed secondary compensation with negative sequence current sharing is activated

Stage 5 (15-16s): Load 2 is disconnected from the SLB

The parameters of the distributed controller are chosen and listed as follows. $K_{P1} = K_{P2} = K_{P3} = K_{P4} = 1$, $K_{I1} = K_{I4} = 10$, $K_{I2} = K_{I3} = 5$, $w_1 : w_2 : w_3 : w_4 = 12 : 6 : 4 : 3$. The simulation results for the 6 stages are shown in Figs. 4.13 and 4.14. As seen from stages 3-5 in Fig. 4.13, the voltage in the SLB is guaranteed to be balanced with the VUF being less than 1% after the compensation at the sacrifice of the unbalanced voltage output in DG1, DG2, DG3, and DG4.

The amplitudes of negative sequence current of each bus are demonstrated, as shown in Fig. 4.14. It is observed that during stage 3, when only SLB compensation is activated, the negative sequence current in SLB is shared by all DGs almost equally. However, during stage 4, when the negative sequence current sharing is activated, each DG shares the negative sequence current according to the designed sharing ratios. This result validates our proposed method.

Acknowledgment

©2017 IEEE. Required, with permission, from Fanghong Guo, Changyun Wen, Jianfeng Mao, and Yong-Duan Song, "Distributed Cooperative Secondary Control for Voltage Unbalance Compensation in an Islanded Microgrid," *IEEE Transactions on Industrial Informatics*, vol. 11, no. 5, pp. 1078 - 1088, 2015.

©2017 IEEE. Required, with permission, from Fanghong Guo, Changyun Wen, Jiawei Chen, "Distributed Voltage Unbalance Compensation in an Islanded Microgrid System by Using Negative Sequence Current Feedback," 2016 14th International Conference on Control, Automation, Robotics and Vision (ICARCV), Phuket, Thailand, 2016.

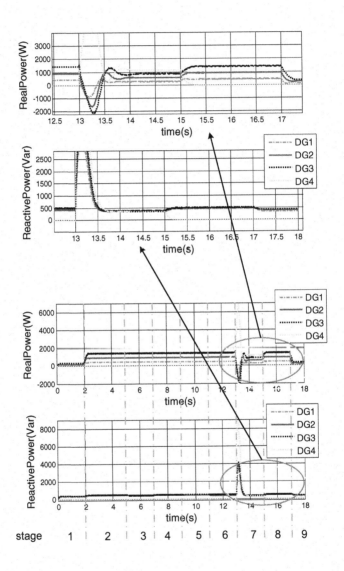

Figure 4.7: Real and reactive power outputs of the MG system

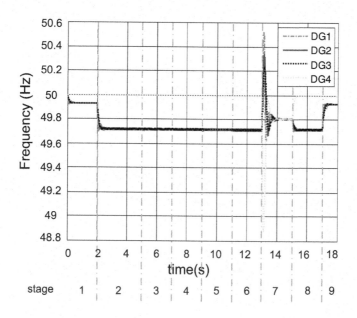

Figure 4.8: Frequency output of the MG system

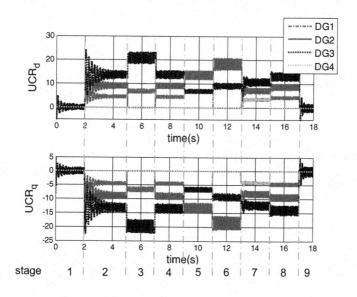

Figure 4.9: The unbalanced compensation references of each DG

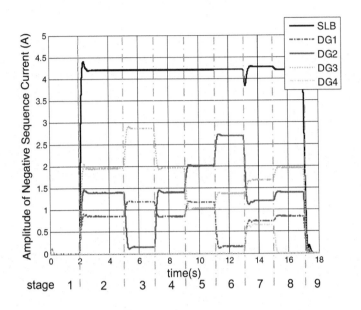

Figure 4.10: Amplitude of negative sequence current

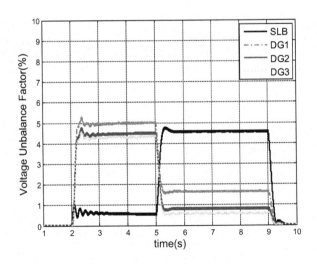

Figure 4.11: Voltage unbalance factors of each DG and SLB in [65]

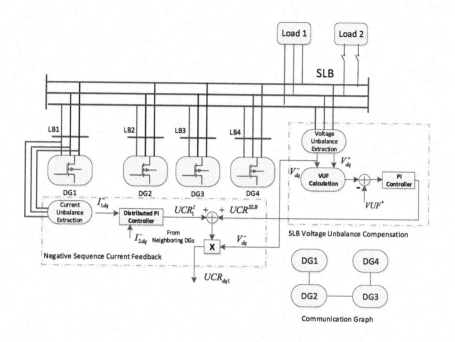

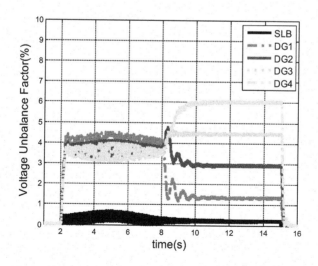

Figure 4.12: Distributed voltage unbalance compensation in an islanded MG using negative sequence current feedback

Figure 4.13: The voltage unbalance factor of each DG and SLB using negative sequence current feedback

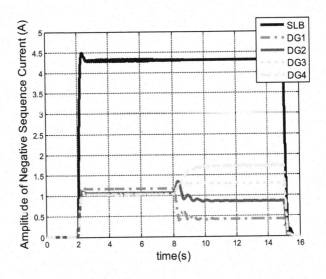

Figure 4.14: Amplitude of negative sequence current using negative sequence current feedback

DISTRIBUTED TERTIARY OPTIMIZATION

Chapter 5

Distributed Single-Area Economic Dispatch

In the last two chapters, distributed control strategies have been successfully applied to solve two key problems in the secondary control layer of MG, i.e., voltage and frequency restoration control, voltage unbalance compensation. In this chapter, we move to the tertiary control layer, and present a distributed economic dispatch (ED) strategy for smart grid systems. Different from the secondary control problems considered in Chapter 3 and 4, in the tertiary layer, we mainly focus on the optimization problem.

Economic dispatch (ED) is considered one of the well-studied and key problems in power system research. It deals with the power allocation among the generators in an economically efficient way while meeting the constraints of total load demand as well as the generator constraints. In this chapter, both conventional thermal generators and wind turbines are taken into account in the economic dispatch model. By decomposing the centralized optimization into optimizations at local agents, a scheme is proposed for each agent to iteratively estimate a solution of the optimization problem in a distributed manner with limited communication among neighbors. It is theoretically shown that the estimated solutions of all the agents reach consensus of the optimal solution asymptotically. This scheme also brings some advantages, such as plug-and-play property. Different from most existing distributed methods, the private confidential information such as gradient or incremental cost of each generator is not required for the information exchange, which makes more sense in real applications. Besides, the proposed method not only handles quadratic but also non-quadratic convex cost functions with arbitrary initial values. Several case studies implemented on a 6-bus power system as well as an IEEE 30-bus power system are discussed and tested to validate the proposed method.

5.1 Introduction

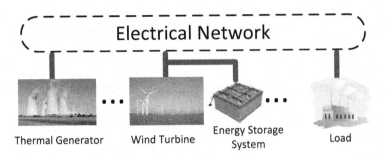

Figure 5.1: Smart grid system

Recently, renewable energy generators have been integrated to power systems to deal with energy and environmental challenges. Among various kinds of renewable energy generators, the wind turbine is widely developed for the advantages such as free availability and environmental friendliness of wind energy as well as maturity of turbine techniques [159], [160]. Hence the ED problem needs to be reformulated not only considering the conventional thermal generators but also the renewable energy generators such as wind turbines. There are mainly two problem formulations and approaches to handle the economic dispatch with random wind power. One is based on the stochastic programming strategies, where only TG cost function is minimized and the wind power is considered as the stochastic constraint appearing in the equality constraint [161], [162]. The other is based on a deterministic model, where the overestimation and underestimation cost of wind power is proposed [163]-[165].

In this chapter, we consider and follow the latter one, where ED for a smart grid system (shown in Fig. 5.1) consisting of conventional thermal generators (TGs), wind turbines (WTs) as well as loads is considered. To ensure high utilization of the intermittent wind power, energy storage systems (ESSs) are always cooperatively integrated with WTs [166]. Note that the quadratic cost function is assumed in most existing ED problem formulations. However, when a WT and ESSs are included, their cost functions are not quadratic any more [163]. Hence some methods mentioned above may fail to work. In this chapter, a distributed ED strategy based on projected gradient and finite-time average consensus algorithm is proposed to solve this new ED problem. Our idea is to let each local agent iteratively estimate a solution of the optimization problem in a distributed manner by using its own and also available information from its neighbors. It is theoretically ensured that the estimated solutions of all the agents converge to the optimal solution of the problem. Several case studies implemented on a 6-bus power system as well as an IEEE 30-bus power system are discussed and tested to validate the proposed method.

Besides the main advantages mentioned earlier in comparing with centralized approaches, our proposed method has some additional advantages over existing distributed schemes, as summarized below:

1. With the proposed method, private confidential information such as gradient or incremental cost is only known by each individual agent and is not used as communication information, which makes more sense in real applications.

2. Compared to λ-consensus algorithm, the cost function is not restricted to be quadratic. Our method can handle ED problems with non-quadratic convex cost functions, such as that of wind turbines. Compared to the distributed gradient method, the initial values of our proposed method can be arbitrary, thus are not required to meet the stringent equality constraint.

5.2 Problem Formulation

Mathematically speaking, the objective of a traditional ED problem is to minimize the total generation cost subject to the demand supply constraint as well as the generator constraints [167]. In this chapter, we consider an ED model which involves random wind power. The main goal of ED is to minimize the system cost consisting of both TGs and WTs, which is given by [163]

$$C(P_i, W_j) = \sum_{i \in S_G} f_i(P_i) + \sum_{j \in S_W} g_j(W_j) \tag{5.1}$$

where S_G and S_W are the sets of TGs and WTs, respectively, P_i, W_j are the power output of the ith TG, $i \in S_G$ and the jth WT, $j \in S_W$.

The cost of conventional TG is usually approximated by a quadratic function [167]:

$$f_i(P_i) = \alpha_i P_i^2 + \beta_i P_i + \gamma_i \tag{5.2}$$

where α_i, β_i, and γ_i are the cost coefficients of the ith TG.

In TG, the scheduled and generated power outputs are always the same. However, due to the random nature of wind speed, the available generated power $W_{j,av}$ at the jth WT is a random variable, which may be different from the scheduled power W_j. Thanks to the integration of ESSs into the WTs, the total output of a WT unit can be guaranteed to be equal to the scheduled one. For example, if the scheduled power output W_j is greater than $W_{j,av}$, then the ESS can compensate the mismatch; if the scheduled power output W_j is less than $W_{j,av}$, then the WT should clamp its output to W_j and the ESSs can be charged by the surplus wind power. In order to characterize the cost of the WT, the overestimation and underestimation cost has been proposed [163], [164].

Referring to [164], the overall cost for j^{th} WT can be expressed as

$$g_j(W_j) = d_j W_j + C_{pwj} E(Y_{ue,j}) + C_{rwj} E(Y_{oe,j}) \tag{5.3}$$

where $d_j W_j$ is a linear cost function for wind power generation with d_j being the cost coefficient or the "price" of the jth WT, the terms $C_{pwj}E(Y_{ue,j})$ and $C_{rwj}E(Y_{oe,j})$ are the underestimation and overestimation costs with C_{pwj}, C_{rwj} being the cost coefficients, respectively, which are explained in detail in the following.

The underestimation cost can be expressed as the penalty cost for not using all the available wind power, which is linear to the mean of random variable $Y_{ue}(= W_{j,av} - W_j)$. The expression of $E(Y_{ue,j})$ is derived in [164] as

$$
\begin{aligned}
E(Y_{ue,j}) \\
= (W_r - W_j) &\left[\exp\left(-\frac{v_r^\kappa}{c^\kappa}\right) - \exp\left(-\frac{v_{out}^\kappa}{c^\kappa}\right) \right] \\
+ &\left(\frac{W_r v_{in}}{v_r - v_{in}} + W_j \right) \left[\exp\left(-\frac{v_r^\kappa}{c^\kappa}\right) - \exp\left(-\frac{v_j^\kappa}{c^\kappa}\right) \right] \\
+ &\frac{W_r c}{v_r - v_{in}} \left\{ \Gamma\left[1 + \frac{1}{\kappa}, \left(\frac{v_j}{c}\right)^\kappa\right] - \Gamma\left[1 + \frac{1}{\kappa}, \left(\frac{v_r}{c}\right)^\kappa\right] \right\}
\end{aligned}
\tag{5.4}
$$

where W_r is the rated wind power, v_r, v_{in}, v_{out} are the rated, cut-in and cut-out wind speeds, κ, c are the *scale factor* and *shape factor* of the Weibull distribution of wind, $\Gamma(a,x)$ is a standard *incomplete gamma function*, v_j is an intermediary variable, which is given as

$$
v_j = v_{in} + \frac{(v_r - v_{in})W_j}{W_r}
\tag{5.5}
$$

Note that in order to simplify the notation, we have dropped the subscript j in the above parameters.

Similarly, the overestimation cost is due to the available wind power being less than the scheduled wind power so it needs to get some power from other source, e.g., ESS, which is expressed as $C_{rwj}E(Y_{oe,j})$, where $E(Y_{oe,j})$ is given as

$$
\begin{aligned}
E(Y_{oe,j}) \\
= W_j &\left[1 - \exp\left(-\frac{v_{in}^\kappa}{c^\kappa}\right) + \exp\left(-\frac{v_{out}^\kappa}{c^\kappa}\right) \right] \\
+ &\left(\frac{W_r v_{in}}{v_r - v_{in}} + W_j \right) \left[\exp\left(-\frac{v_{in}^\kappa}{c^\kappa}\right) - \exp\left(-\frac{v_j^\kappa}{c^\kappa}\right) \right] \\
+ &\frac{W_r c}{v_r - v_{in}} \left\{ \Gamma\left[1 + \frac{1}{\kappa}, \left(\frac{v_j}{c}\right)^\kappa\right] - \Gamma\left[1 + \frac{1}{\kappa}, \left(\frac{v_{in}}{c}\right)^\kappa\right] \right\}
\end{aligned}
\tag{5.6}
$$

Then considering both generator constraints and demand supply constraint, the ED problem with random wind power can be formulated as

$$
\left\{
\begin{aligned}
&\min_{P_i, W_j} C(P_i, W_j) \\
&s.t. \ \sum_{i \in S_G} P_i + \sum_{j \in S_W} W_j = P_d \\
&\quad P_i^{\min} \le P_i \le P_i^{\max}, i \in S_G \\
&\quad 0 \le W_j \le W_{r,j}, j \in S_W
\end{aligned}
\right.
\tag{5.7}
$$

where $P_i^{\min}$, $P_i^{\max}$ are the lower and upper bounds of the ith TG, $W_{r,j}$ is the rated wind power of the jth WT, P_d is the total load demand satisfying $\sum_{i \in S_G} P_i^{\min} \leq P_d \leq \sum_{i \in S_G} P_i^{\max} + \sum_{j \in S_W} W_{r,j}$.

Note that the cost coefficient α_i of the TG is usually positive, which implies that Eqn. (5.2) is a convex function. Meanwhile, it is also proved in [164] that Eqn. (5.4) and (5.6) are also convex with respect to W_j, which yields the convexity of the WT cost function (5.1). Also the constraints are convex, thus the ED problem described in (5.7) can be considered as a convex optimization problem.

Remark 5.1 *Compared to the ED problem formulated in [85]-[91], the cost function in (5.7) is not quadratic any more, which implies that the designed distributed methods in [85]-[88] may fail to work. In this chapter, a new distributed optimization method will be introduced to solve the ED problem formulated in (5.7).*

Suppose there are $N = n_G + n_W$ generators, consisting of $n_G = |S_G|$ TGs and $n_W = |S_W|$ WTs, and M loads in a smart grid system, shown in Fig. 5.1. We first treat every generator and load as an "agent," and each agent is assigned a unique ID. Without loss of generality, we assign the first N agents as the TGs and WTs and denote their estimated generated power in a global vector as $x = \begin{bmatrix} x_1 & \cdots & x_{n_G} & \cdots & x_N \end{bmatrix}^T$. The cost function c_k and constraint set X_k of agent $k, k \in S_G \cup S_W$ are denoted as follows respectively.

$$c_k(x) = \begin{cases} f_k(x_k), k \in S_G, k = 1, \cdots, n_G \\ g_k(x_k), k \in S_W, k = n_G + 1, \cdots, N \end{cases} \tag{5.8}$$

$$X_k = \begin{cases} \begin{aligned} P_k^{\min} \leq x_k \leq P_k^{\max} \\ \sum_{i=1}^{N} x_i = P_d \end{aligned} \quad , k \in S_G, k = 1, \cdots, n_G \\ \begin{aligned} 0 \leq x_k \leq W_{r,k} \\ \sum_{i=1}^{N} x_i = P_d \end{aligned} \quad , k \in S_W, k = n_G + 1, \cdots, N \end{cases} \tag{5.9}$$

Then (5.7) can be reformulated as

$$\begin{cases} \min_x \sum_{k=1}^{N} c_k(x) \\ s.t. \quad x \in \cap_{k=1}^{N} X_k \end{cases} \tag{5.10}$$

Lemma 5.1 *The ED problem formulated in (5.10) has an optimal solution x^*.*

Proof : As the constraint set X_k in (5.9) is compact, thus the intersection $X = \cap_{k=1}^{N} X_k$ is also compact. Besides, the function $c_k(x)$ is a continuous convex function in $\mathbb{R}^{N \times 1}$, which implies that $\sum_{k=1}^{N} c_k(x)$ is also a continuous function. It follows from the well-known *extreme value theorem* that the ED problem formulated in (5.10) has an optimal solution $x^*, x^* \in \cap_{k=1}^{N} X_k$. ■

Note that the optimal solution x^* is unknown to each agent and both cost function $c_k(x)$ and constraint X_k of generator k, $k \in S_G \cup S_W$ are only known by agent k itself. Our idea is that each agent estimates the optimal solution by using the available information of its neighboring agents and itself iteratively. Denoting the estimate of kth agent at time step l as $x^k(l)$, then our aim is to propose a scheme to achieve that $\lim_{l \to \infty} x^k(l) = x^*$, $k = 1, \cdots, N$. In other words, the estimates of all agents reach consensus of the optimal solution asymptotically.

5.3   Total Load Demand Discovery

In this section, a distributed load demand discovery method is introduced to determine P_d in the constraint (5.9) by applying the modified FACA algorithm.

Note that the total load demand P_d appears in the constraint set (5.9) of each agent. How to determine P_d in a distributed way is another problem to be handled in this section. Here we propose to apply distributed FACA to find P_d for each agent. Let y^i be the communication state for the ith agent, and its initial value is defined as

$$y^i(0) = \begin{cases} 0, i \subset S_G \cup S_W, & i = 1, \cdots, N \\ P_d^i, i \in S_L, & i = N+1, \cdots, N+M \end{cases} \tag{5.11}$$

Lemma 5.2 *The total load demand P_d can be determined by each agent i, $i \in S_G \cup S_W$ in finite K steps when using the FACA updating law (2.1) and (2.2), where K is defined in Lemma 2.1 in Chapter 2.*

Proof :   According to Lemma 2.1 , $y^i(m), i = 1, \cdots, N+M$ will reach an average consensus in K steps, *namely,* $y^i(K) = y^j(K) = \frac{\sum_{i=1}^{N+M} y^i(0)}{N+M}, \forall i, j = 1, \cdots, N+M$. Then the total load demand P_d can be obtained by each agent i, $i \in S_G \cup S_W$, i.e.,

$$P_d = (N+M)y^i(K), \quad i = 1, \cdots, N. \tag{5.12}$$

■

An illustration example is shown in Table 5.1. In this example, agents 1 and 2 are the generators while agents 3 and 4 are loads with the demand of 3 and 5 respectively. At $k = 0$, each agent sets their initial value according to (5.11) as $y^1(0) = 0, y^2(0) = 0, y^3(0) = 3, y^4(0) = 5$ respectively. By applying distributed FACA, they can reach a consensus $y^1(3) = y^2(3) = y^3(3) = y^4(3) = 2$ in 3 steps and the total load demand P_d is obtained as $P_d = 4*2 = 8$.

Table 5.1: Illustration example of total load demand discovery

Step	Content	Communication Graph
$m = 0$	$y^1(0) = 0$ $y^2(0) = 0$ $y^3(0) = 3$ $y^4(0) = 5$	
$m = 1$	$y^1(1) = 0$ $y^2(1) = 5.1212$ $y^3(1) = 1.2929$ $y^4(1) = 1.5859$	
$m = 2$	$y^1(2) = 2.5606$ $y^2(2) = 0.6465$ $y^3(2) = 3.3535$ $y^4(2) = 1.4394$	
$m = 3$	$y^1(3) = 2$ $y^2(3) = 2$ $y^3(3) = 2$ $y^4(3) = 2$	

$$\mathcal{L} = \begin{bmatrix} 1 & -1 & 0 & 0 \\ -1 & 2 & -1 & 0 \\ 0 & -1 & 2 & -1 \\ 0 & 0 & -1 & 1 \end{bmatrix}$$

5.4 Distributed Economic Dispatch

5.4.1 *Distributed projected gradient method (DPGM)*

Motivated by the projection idea in [168] and constrained optimization in [130], we now propose a distributed projected gradient method to solve the ED problem formulated in (5.10). Different from existing distributed methods in [85]-[88], the communication information required here is the estimates of the scheduled generated power x^k, i.e., the estimated optimal solutions, of the local agent and its neighbors rather than the more private and confidential gradient or incremental gain. Recall that $x^k(l)$, $k = 1, \cdots, N + M$, denotes the estimate of the agent k at iteration l, which is an $N \times 1$ vector. As only generator agents estimate the power output, we define $x^k(l) = \begin{bmatrix} 0 & \cdots & 0 \end{bmatrix}^T$, $k = N + 1, \cdots, N + M$ for all load agents for any l. Unlike the distributed gradient method in [89]-[91], the initial value $x^k(0)$, $k = 1, \cdots, N$ is allowed to be arbitrary.

The kth agent updates its estimate by using the average information produced by distributed FACA, then taking a gradient step to minimize its own cost function c_k, and at last projecting the result on its constraint set X_k. This updating rule can be summarized as

$$
\begin{cases}
z_1^k(l) = w_{kk}(1)x^k(l) + \sum_{j \in \mathcal{N}_k} w_{kj}(1)x^j(l) \\
z_2^k(l) = w_{kk}(2)z_1^k(l) + \sum_{j \in \mathcal{N}_k} w_{kj}(2)z_1^j(l) \\
\quad\vdots \\
z_K^k(l) = w_{kk}(K)z_{K-1}^k(l) + \sum_{j \in \mathcal{N}_k} w_{kj}(K)z_{K-1}^j(l) \\
z^k(l) = \frac{N+M}{N}z_K^k(l)
\end{cases}
\tag{5.13}
$$

$$
x^k(l+1) = P_{X_k}\left[z^k(l) - \zeta_l \nabla c_k(z^k(l))\right]
\tag{5.14}
$$

where $w_{kk}(m)$, $w_{kj}(m)$, $m = 1, \cdots, K$ are the FACA updating gains determined in (2.1), $P_{X_k}[.]$ is the projection operator described in Section 2.4.3, ζ_l is a stepsize at iteration l, ∇c_k denotes the gradient of the cost function c_k.

Theorem 5.1 *Let $\{x^k(l)\}$, $k = 1, \cdots, N$ be the estimates generated by the algorithm (5.13)-(5.14) and $X = \cap_{k=1}^N X_k$ be the intersection set. Then the sequence $\{x^k(l)\}$, $k = 1, \cdots, N$ converges to the optimal solution $x^\star$ with $x^\star \in X$, i.e.,*

$$
\lim_{l \to \infty} x^k(l) = x^\star, \; k = 1, \cdots, N
$$

if the stepsize ζ_l satisfies that $\zeta_l > 0, \sum_l \zeta_l = \infty$ and $\sum_l \zeta_l^2 < \infty$.

Proof : According to Lemma 2.1, the average consensus process (5.13) can be reached in finite K steps if the update gains $w_{ii}(m)$ and $w_{ij}(m)$, $m = 1, \cdots, K$ are chosen as (2.2). This process is equal to $z^k(l) = \frac{1}{N}\sum_{j=1}^N x^j(l)$. Then the result follows from the proof of *Proposition 5* in [130]. Due to the page limit, we omit the details here. ■

Remark 5.2 *Compared to the projected subgradient algorithm in [130], our proposed DPGM is a fully distributed one. The average step (5.13) in [130] requires that each agent communicates with all the others in the entire system, which almost has the same communication cost as a centralized one. By applying the distributed FACA algorithm, the proposed DPGM can realize the same function with limited communication.*

Remark 5.3 *Compared to the existing methods for the ED problem, the proposed DPGM has the following advantages: 1) No private information such as gradient or the incremental cost is required to exchange with other generators; 2) It can solve any convex objective cost function rather than only quadratic function; 3) The initial value of the estimate can be chosen arbitrarily by each agent individually.*

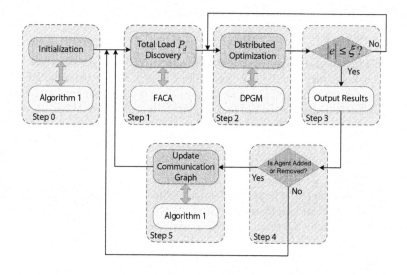

Figure 5.2: Flowchart of distributed economic dispatch

5.4.2 Implementation of distributed ED

Based on proposed Algorithm 2.1, distributed FACA, and DPGM, we are ready to implement our distributed ED design. The flowchart of our proposed distributed ED procedure is shown in Fig. 5.2, with each corresponding step described as follows:

Step 0: Initialization and graph discovery: As a starting point, the communication graph of all agents is pre-designed as connected.[1] Using Algorithm 1, each agent can get the information of the whole communication graph such as the number of generator agents N, the number of load agents M, and the Laplacian matrix $\mathcal{L}$ in less than $N - 1$ steps, then using some available numerical methods to calculate the nonzero eigenvalues of $\mathcal{L}$.

Step 1: Total load demand discovery: In this step, each agent determines the total load demand P_d from (5.12) in K steps according to Lemma 5.2.

Step 2: Distributed optimization: The DPGM algorithm (5.13)-(5.14) introduced in Section 4.1 is applied here to execute the distributed ED.

Step 3: Stop criterion check: In theory, the sequences of estimates generated by DPGM converge to the optimal solution asymptomatically. In practice, some iteration stop criterions are set. Here we set $|e^k| \leq \xi, k = 1, \cdots, N$ as a stop criterion, where $e^k = x^k(l+1) - x^k(l)$, ξ is a user-defined small positive number.

---

[1]This is easily implementable, as each agent can choose the same communication graph as its physical connection graph at the initial step.

If this condition is satisfied, then stop the iteration, output the results, and go to Step 4. If not, go to Step 2.

Step 4: Plug-in and plug-out reconfiguration: At this step, each agent needs to check whether there is any agent added to or removed from the grid. If yes, execute the *Graph Reconfiguration* rule described in Chapter 4, and then go to Step 5 to update the communication graph; otherwise go to Step 1.

Step 5: Graph updating: At this step, all agents need to update the communication graph using Algorithm 2.1 and then go to Step 1.

A simple graph reconfiguration example is illustrated in Fig. 5.3. Suppose agent 3 is removed, and its neighboring agent $\mathcal{N}_3 = \{1,2,4\}$ monitors this situation, respectively. For agent 1, it needs to set up a new communication channel with agent 4 (no need with agent 2 as they have already been connected); for agent 2, it also needs to set up a new communication channel with agent 4; for agent 4, it needs to set up communication with both agent 1 and 2.

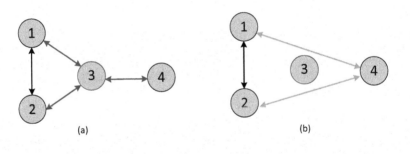

(a)   (b)

Figure 5.3: Graph reconfiguration illustration example

5.4.3 Complexity analysis

Here we analyze the computational performance of proposed distributed strategy. Note that our proposed DPGM (5.13)-(5.14) mainly contains two parts, *namely*, finite-time average consensus (FAC) and projected gradient operation (PGO). According to Lemma 2.1, for a system with N agents, the FAC process can be fulfilled in less than $N - 1$ steps, which is much more efficient than a conventional average consensus algorithm possessing asymptotical convergence. For PGO, it is actually a combination of the gradient descent method and a projection operation. The projection operation in our specific problem, as discussed in Section 2.4.3 in Chapter 2, is nothing but a simple algebraic operation. So it has little contribution to the computational cost.

5.5 Case Studies

In order to test the effectiveness of the proposed distributed ED method, several case studies are presented and discussed in this section. Firstly, a 6-bus power system implementation without and with generator constraints is demonstrated. The second case study illustrates the plug-and-play property of the proposed method including both generator and load node. Then the IEEE 30-bus system is used as a large network case to demonstrate the effectiveness of the proposed method. Lastly, comparisons with the genetic algorithm are carried out.

5.5.1 Case study 1: Implementation on 6-bus power system

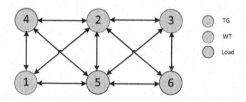

Figure 5.4: The communication graph of test power system

In this test case, a 6-bus power system topology is adopted from [87]. It consists of 3 TGs, 1 WT, and 2 load nodes. We replace 1 TG with a WT in [87]. Its communication graph is shown in Fig. 5.4. The corresponding Laplacian matrix can be obtained by each agent using Algorithm 2.1, which is

$$
\mathcal{L} =
\begin{bmatrix}
3 & -1 & 0 & -1 & -1 & 0 \\
-1 & 5 & -1 & -1 & -1 & -1 \\
0 & -1 & 3 & 0 & -1 & -1 \\
-1 & -1 & 0 & 3 & -1 & 0 \\
-1 & -1 & -1 & -1 & 5 & -1 \\
0 & -1 & -1 & 0 & -1 & 3
\end{bmatrix}
$$

Its 3 distinct nonzero eigenvalues are $\lambda_1 = 2, \lambda_2 = 4, \lambda_3 = 6$, which means that it needs only $K = 3$ steps for each agent to reach consensus when applying distributed FACA. The parameters of three different types of TGs are adopted from [87], while the WT parameters are from [164]. These parameters are listed in Tables 5.2 and 5.3, respectively.

Firstly, the generator constraints are not imposed. In the initial, the load demand is $P_d^5 = 200MW$, $P_d^6 = 200MW$, and each generator is operating in the optimal condition with the generated power $P_1 = 174.0683MW$, $P_2 = 100.00MW$, $P_3 = 50.00MW$, $W_1 = 75.9317MW$. Then load 5 is doubled, i.e., $P_d^5 = 400MW$. The

Table 5.2: Parameters of thermal generators

Generator	α_i [$/MW^2h$]	β_i [$/MWh$]	γ_i [$/h$]	$P_i^{\min}$ [MW]	$P_i^{\max}$ [MW]
1	0.00142	7.2	510	150	600
2	0.00194	7.85	310	100	400
3	0.00482	7.97	78	50	200

Table 5.3: Parameters of wind and wind turbine

Wind	v_{in}	v_{out}	v_r	(c,k)
	5	45	15	(8,2)

Wind Turbine	d_j	C_{pwj}	C_{rwj}	W_r
	6	3.1	3.1	160

changed total load demand can be discovered by TG and WT generators in 3 steps as shown in Table 5.4. Then each generator agent (Agents 1 -4) conducts the distributed ED using the proposed DPGM method. The initial value is chosen as $x^1(0) = \begin{bmatrix} 174.0683 & 0 & 0 & 0 \end{bmatrix}^T$, $x^2(0) = \begin{bmatrix} 0 & 100 & 0 & 0 \end{bmatrix}^T$, $x^3(0) = \begin{bmatrix} 0 & 0 & 50 & 0 \end{bmatrix}^T$, $x^4(0) = \begin{bmatrix} 0 & 0 & 0 & 75.9318 \end{bmatrix}^T$. The iteration process is shown in Fig. 5.5 (a). Clearly all the estimated power outputs converge to the optimized solution $P_1^\star = 367.7996MW$, $P_2^\star = 102.2463MW$, $P_3^\star = 29.1174MW$, $W_1^\star = 100.8367MW$ with a total cost $C = 5611.8\$$. Also it is found that the incremental cost (or the gradient of the cost ∇c_k) of each generator reaches a consensus value 8.25. Note that the third generator is in conflict with its lower output bound $P_3^{\min} - 50MW$.

Table 5.4: Total load demand discovery

Step	Content	Step	Content
$m = 0$	$y^1(0) = 0$ $y^2(0) = 0$ $y^3(0) = 0$ $y^4(0) = 0$ $y^5(0) = 400$ $y^6(0) = 200$	$m = 2$	$y^1(2) = 50$ $y^2(2) = 0$ $y^3(2) = 50$ $y^4(2) = 50$ $y^5(2) = 400$ $y^6(2) = 50$
$m = 1$	$y^1(1) = 200$ $y^2(1) = 300$ $y^3(1) = 300$ $y^4(1) = 200$ $y^5(1) = -500$ $y^6(1) = 100$	$m = 3$	$y^1(3) = 100$ $y^2(3) = 100$ $y^3(3) = 100$ $y^4(3) = 100$ $y^5(3) = 100$ $y^6(3) = 100$

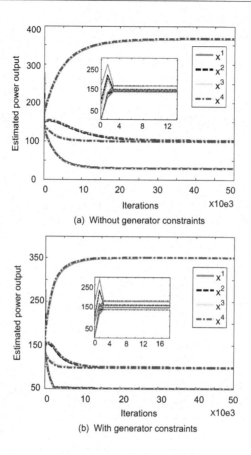

(a) Without generator constraints

(b) With generator constraints

Figure 5.5: Simulation results of 6-bus power system

Then, we consider the generator constraints. The results are shown in Fig. 5.5 (b). The optimized power output are $P_1^\star = 351.7814MW$, $P_2^\star = 100.0248MW$, $P_3^\star = 50.0248MW$, $W_1^\star = 98.1691MW$ with a total cost $C = 5614.4\$$. In this case, note that all the generators' power outputs are within their constraints.

This case study shows that our proposed method can handle the ED problem both without and with generator constraints.

5.5.2 Case study 2: Plug-and-play capability

This case study is to test the flexibility of the proposed method. The plug-and-play performance of both generator and load are considered.

5.5.2.1 Generator plug-and-play

In this subcase, the plug-and-play of TG is considered. The results of TG are shown in Fig. 5.6. From Fig. 5.6 (a), it is clear that the estimated power outputs of all agents $x^k(l)$, $k = 1, \cdots, 4$ almost reach consensus in $l = 10 \times 10^3$ iterations. To clearly demonstrate the estimated power output, the estimates of agent 1 is shown in Fig. 5.6 (b). In the initial, the load demands are $P_d{}^5 = 400MW$ and $P_d{}^6 = 200MW$. After a few iterations the proposed method ensures convergence to the optimized power output $P_1{}^* = 351.7814MW$, $P_2{}^* = 100.0248MW$, $P_3{}^* = 50.0248MW$, $W_1{}^* = 98.1691MW$ with the total cost $C = 5614.4\$$. Suppose generator 1 (agent 1) is disconnected from the power system at the time step $l = 30 \times 10^3$. Then using the *Graph Reconfiguration* rule introduced in Section 4.2, each agent can reconfigure the communication graph as shown in Fig. 5.7. The remaining generators can converge to new optimized power output $P_1{}^* = 0MW$, $P_2{}^* = 322.2944MW$, $P_3{}^* = 117.6226MW$, $W_1{}^* = 160.0829MW$. It is shown in Fig. 5.6 (c) that the total cost increases to $C = 5960.7\$$ due to the absence of generator 1. It further indicates that generator 1 is more economically efficient compared to the average of other generators. Then at the time step $l = 60 \times 10^3$, generator 1 is connected to the system again and all the results converge to those of the previous ones.

5.5.2.2 Load plug-and-play

The results of this subcase are shown in Fig. 5.8. The initial condition is the same as that in subcase 1). At the time step $l = 30 \times 10^3$, load 6 (agent 6) is disconnected from the power system. The other agents detect this change, update the communication graph as well as the total load demand $P_d = 400MW$. Then the remaining agents ensure the convergence to new optimized power outputs $P_1{}^* = 173.56MW$, $P_2{}^* = 100.0785MW$, $P_3{}^* = 50.0784MW$, $W_1{}^* = 76.2811MW$ with the total cost $C = 4024.7\$$. At the time step $l = 60 \times 10^3$, load 6 is connected to the system again and all the values are back to the previous values before load 6 is disconnected.

Both subcases 1) and 2) show that the proposed method is fully distributed and has the plug-and-play property.

5.5.3 Case study 3: Implementation on IEEE 30-bus test system

In order to test the effectiveness of the proposed method for a large network, the IEEE 30-bus system is chosen as a test system. The generator and load bus parameters are adopted from [90], which are also listed in Table 5.5 and Table 5.6, respectively. First, the communication graph can be chosen to be the same as the physical connections. Then the total load demand $P_d = \sum_{i=1}^{30} P_i^d = 283.4MW$ can be easily discovered by applying proposed distributed FACA. The optimized power allocation can be obtained by applying DPGM. Suppose that at the time steps $l = 30 \times 10^3$ and $l = 60 \times 10^3$ the load demand is increased by 30% and reduced by 20%, respectively. The simulation results are shown in Fig. 5.9 and Table 5.7. These simulation results illustrate the effectiveness of the proposed method applied on a large network.

Table 5.5: Generator parameters in IEEE 30-bus

Generator No.	α_i $[\$/MW^2h]$	β_i $[\$/MWh]$	γ_i $[\$/h]$	P_i^{min} $[MW]$	P_i^{max} $[MW]$
1	0.00375	2	0	50	200
2	0.0175	1.75	0	20	80
5	0.0625	1.0	0	15	50
8	0.00834	3.25	0	10	35
11	0.025	3.0	0	10	30
13	0.025	3.0	0	12	40

Table 5.6: Load parameters in IEEE 30-bus

Load No.	P_i^d $[MW]$	Load No.	P_i^d $[MW]$	Load No.	P_i^d $[MW]$
1	0.00	11	0.00	21	17.5
2	21.7	12	11.2	22	0.00
3	2.40	13	0.00	23	3.20
4	7.60	14	6.20	24	8.70
5	94.2	15	8.20	25	0.00
6	0.00	16	3.50	26	3.50
7	22.8	17	9.00	27	0.00
8	30.0	18	3.20	28	0.00
9	0.00	19	9.50	29	2.40
10	5.80	20	2.20	30	10.6

5.5.4   Case study 4: Comparison with heuristic search method

There are various kinds of heuristic search methods applied to solve the conventional ED problems, such as genetic algorithm (GA) [169], modified GA [170], particle swarm optimization [171], and the Monte Carlo method [172]. In this section, we apply one popular heuristic search method, i.e., GA method in [169], to our proposed ED problem for comparison. The parameters of the test system are the same as those in case study 1. Some key parameters of the GA method are listed in Table 5.8. The fitness function is chosen as the total cost in Eqn. (5.1) plus some penalty

Table 5.7: Simulation results of IEEE 30-bus test system

Total Demand P_d [MW]	Optimized Power Output P_i [MW]					
	Gen 1	Gen 2	Gen 5	Gen 8	Gen 11	Gen 13
283.4	185.03	47.11	19.20	10.02	10.02	12.02
368.42	200.03	66.13	24.57	35.03	21.33	21.33
226.72	140.71	37.49	16.52	10.00	10.00	12.00

functions constraining the variables according to [169]. The evolution of the fitness value is shown in Fig. 5.10. The best fitness value for a population is the smallest fitness value among all the individuals in the population, while the mean fitness value is the average of their fitness values [169]. After 51 generations, the mean fitness approaches the best fitness value. The GA stops when the average relative change in the fitness value is less than the pre-defined function tolerance, which is listed in Table 5.8. The final power outputs are obtained from the best individual of the last generation as $P_1^\star = 349.06MW$, $P_2^\star = 103.5257MW$, $P_3^\star = 50.0306MW$, $W_1^\star = 97.3836MW$ with the total cost $C = 5614.7\$$. Comparing to the results in case study 1 ($P_1^\star = 351.7814MW, P_2^\star = 100.0248MW, P_3^\star = 50.0248MW, W_1^\star = 98.1691MW$ with a total cast $C = 5614.4\$$), they are about the same. The consistency of these results further illustrates and verifies the effectiveness of our scheme.

Table 5.8: Parameters of GA method

Parameter	Valuc	Parameter	Value
Population Size	80	Crossover Fraction	0.8
Migration Interval	20	Migration Interval Fraction	0.2
Maximum Generations	60	Penalty Factor	100
Stall Time Limit	200	Function Tolerance	1e-6

However, such similar results are achieved with three major differences between the GA method and our proposed approach. Firstly, similar to most heuristic search methods, the GA method is a centralized one while ours is fully distributed. Secondly, the GA method does not have the plug-and-play property. Thirdly, the operation and implementation costs are different.

Acknowledgment

©2017 IEEE. Required, with permission, from Fanghong Guo, Changyun Wen, Jianfeng Mao, and Yong-Duan Song, "Distributed Economic Dispatch for Smart Grids with Random Wind Power," *IEEE Transactions on Smart Grid*, vol. 7, no. 3, pp. 1572 -1583, 2016.

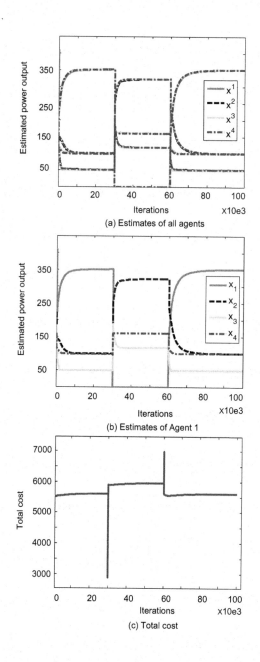

(a) Estimates of all agents

(b) Estimates of Agent 1

(c) Total cost

Figure 5.6: Simulation results with generator plug-and-play

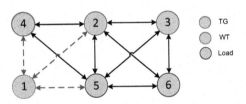

Figure 5.7: Graph reconfiguration of generator 1 plug-and-play

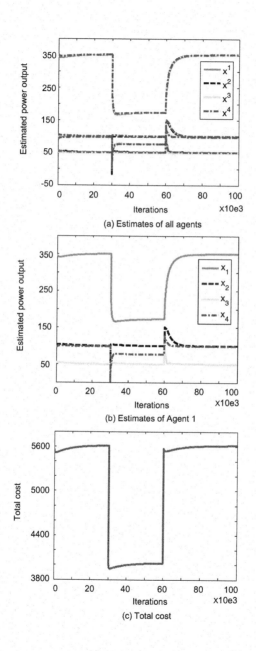

(a) Estimates of all agents

(b) Estimates of Agent 1

(c) Total cost

Figure 5.8: Simulation results with load plug-and-play

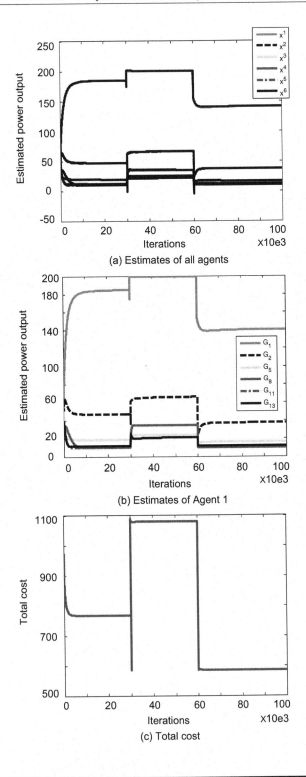

(a) Estimates of all agents

(b) Estimates of Agent 1

(c) Total cost

Figure 5.9: Simulation results of IEEE 30-bus system

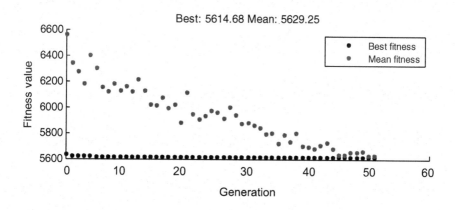

Figure 5.10: Evolution of fitness values in GA method

Chapter 6

Distributed Multi-Area Economic Dispatch

In Chapter 5, an economic dispatch method for one single area power system is considered. In this chapter, the economic dispatch for a multi-area power system is carried out. Firstly, two distributed multi-cluster optimization algorithms are proposed. Then we prove the convergence of the proposed two multi-cluster algorithms. Lastly, we apply the proposed algorithms to solve the multi-area ED problem.

6.1 Introduction

Recently, due to rapid growth of electricity markets, the electrical grid has become an interconnected large-scale system. In this scenario, if we apply the distributed ED method in Chapter 5, it would take many more communication steps just for one internal iteration, which would be quite time-consuming. In order to overcome this drawback, in this chapter, we first propose to divide such a large-scale system into several clusters and each cluster has a leader to communicate with the leaders of its neighboring clusters. The agents in the same cluster conduct local optimization and communicate with their neighboring agents synchronously by applying FACA to reach a cluster average consensus. Then two interesting distributed optimization algorithms are proposed based on the different communication strategies for leader agents, i.e., synchronous and sequential algorithms, as shown in Fig. 6.1. In the synchronous algorithm, the leader agent communicates with the leader agents in its neighboring cluster to reach a consensus of a global average estimate. In the next iteration, each agent conducts its local optimization based on this global average estimate. While in sequential algorithm, the leader agent passes the cluster estimate to a leader agent in its neighboring cluster. In one iteration, each cluster is ensured to have one chance to update its estimate.

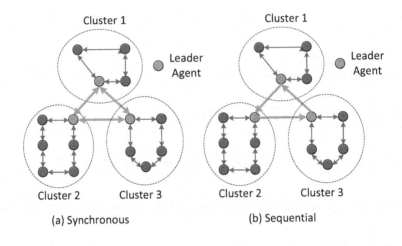

Figure 6.1: Two different communication strategies

It is worthy to point out that in both algorithms, an extra manipulation is conducted by each leader agent. Such extra manipulation can be regarded as the effect of *virtual agents,* a new idea to be proposed to achieve convergence property. With the help of virtual agent, we theoretically establish the convergence property for these two proposed algorithms.

Comparing these proposed algorithms, each algorithm has its own advantages. The synchronous algorithm allows the leader agent exchange their estimate simultaneously and all the cluster conduct optimization in parallel, which takes less time in one iteration. However, when the agents in the system are sparsely distributed, distributed optimization with synchronous communication may not be proper. For example, for a sensor network system, the sensors are located sparsely, then their communication latencies are quite different when using the synchronous communication strategy. In this scenario, the sequential algorithm is more proper.

6.2 Problem Formulation

Motivated by the optimization problems in practical systems such as power system [117], [118], and wireless network system [119], a large-scale multi-agent system with $m = |\mathcal{M}|$ clusters is considered; $\mathcal{M}$ is the set of clusters. Each cluster has $n_i = |\mathcal{A}_i|$ agents, with $\mathcal{A}_i, \forall i = 1, \cdots, m$ denoting the set of agents in the ith cluster. Each agent j in cluster i has its local objective function $f_i^j(x)$ and local constraint set X_i^j, which are only known to agent j itself and cannot be shared with other agents. Also a global constraint set X_g is imposed and known to all the agents. The goal of the agents is to cooperatively solve the constrained optimization problem

$$\begin{cases} \min_x \sum_{i=1}^{m} \sum_{j=1}^{n_i} f_i^j(x) \\ s.t. \ x \in \bigcap_{i=1}^{m} \bigcap_{j=1}^{n_i} X_i^j \cap X_g \end{cases} \tag{6.1}$$

where $f_i^j(x) : \mathbb{R}^n \to \mathbb{R}$ and X_i^j, $i = 1, \cdots, m$, $j = 1, \cdots, n_i$ are a convex function and compact convex sets, respectively.

To obtain a more compact formulation, we merge the global constraint set X_g into the local set X_i^j and generate a new constraint set $\bar{X}_i^j = X_i^j \cap X_g$, $\forall i \in \mathcal{M}$, $j \in \mathcal{A}_i$. Then we can reformulate (6.1) as

$$\begin{cases} \min_x \sum_{i=1}^{m} \sum_{j=1}^{n_i} f_i^j(x) \\ s.t. \ x \in X \end{cases} \tag{6.2}$$

where X is the intersection set of all the local constraint sets, i.e., $X = \bigcap_{i=1}^{m} \bigcap_{j=1}^{n_i} \bar{X}_i^j$.

Denote the optimal solution of (6.2) as $x^\star \in X$; it is easy to conclude that $x^\star$ exists according to the well-known *extreme value theorem* [173]. But it is unknown to each agent. Our idea is to let each agent estimate the optimal solution by using the available information of its neighboring agents and itself iteratively. Denote the estimate of the agent j in group i at iteration l as $\hat{x}_i^j(l)$, $i = 1, \cdots, m$, $j = 1, \cdots, n_i$. Then our objective is to propose an algorithm to ensure all these estimates reach consensus of the optimal solution as iterations increase, i.e, $\lim_{l \to \infty} \hat{x}_i^j(l) = x^\star$, for all $i = 1, \cdots, m$, $j = 1, \cdots, n_i$.

To achieve this, we make the following assumptions.

Assumption 6.1 *Function f_i^j is convex and differentiable.*

Suppose ∇f_i^j is the gradient function of f_i^j, then from [132] and [133], Assumption 6.1 ensures that it is bounded over the set X, i.e., there exists a scalar $L > 0$, such that

$$\left\| \nabla f_i^j(x) \right\| \leq L, \quad \forall x \in X \tag{6.3}$$

if the set X is compact [131].

Remark 6.1 *In this chapter, we assume f_i^j is convex and differentiable, so that its gradient $\nabla f_i^j(x)$ exists for any $x \in \mathbb{R}^n$. However, this assumption can be relaxed so that f_i^j is only convex and allowed to be non-differentiable at some points. In this case, a **subgradient** exists and can be used in the role of a gradient [130].*

6.3 Distributed Optimization Algorithm

In this section, two distributed synchronous optimization algorithms are proposed to solve the problem formulated in (6.2). First, we assign one agent as leader agent in each cluster. Without loss of generality, the leader agent is labeled as Agent 1 in each cluster.

To solve the proposed optimization problem, we make the following assumptions regarding the communication graph.

Assumption 6.2 *The communication graph among the agents in the same cluster i,* $\mathcal{G}_i = (V_i, \xi_i)$ *is undirected and connected.*

Assumption 6.3 *The communication graph among the leader agents* $\mathcal{G}_{leader} = (\mathcal{V}_{leader}, \xi_{leader})$ *is connected.*

6.3.1 Distributed synchronous optimization algorithm

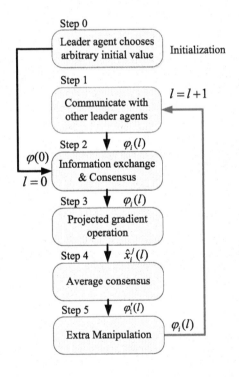

Figure 6.2: Flowchart of distributed synchronous optimization

The flowchart of the proposed distributed optimization method is shown in Fig. 6.2, with each corresponding step described as follows:

Step 0: Initialization: As a starting point, at $l = 0$, the leader agent in each cluster starts estimating the optimal solution by choosing one arbitrary initial value $\varphi(0) \in \mathbb{R}^n$ and then goes to Step 2.

Step 1: Leader agents average consensus: Suppose at iteration step l, $l \geq 1$, the leader agent in Cluster i, $i \in \mathcal{M}$ communicates with the leader agents in its neighboring clusters $k, k \in \mathcal{N}_i$, where $\mathcal{N}_i$ denotes the neighborhood set of the area m. Initially, each leader agent sets the initial value as $z_i^0 = \varphi_i(l-1)$, where $\varphi_i(l-1)$ is called *modified cluster estimate*, which is obtained in Step 5 later. By applying the FACA algorithm, after K' steps, each leader agent reaches average consensus so that

$$\varphi(l) = z_i^{K'} = \frac{\sum\limits_{i=1}^{m} z_i^0}{m} = \frac{\sum\limits_{i=1}^{m} \varphi_i(l-1)}{m} \qquad (6.4)$$

where $\varphi(l)$ is called *average estimate*, K' is the number of distinct nonzero eigenvalues of the Laplacian matrix of the communication graph among the leader agents.

Step 2: Information exchange and consensus: Within Cluster i, by letting each agent exchange information with its neighbors, all the agents reach an average consensus of the *average estimate* $\varphi(l)$. This process can be summarized as follows. According to Lemma 2.1, for $l \geq 1$, the initial value can be set as $y_1^0 = \varphi(l), y_2^0 = \cdots = y_{n_i+q_i}^0 = 0$. Then by using FACA, after K_i steps, all the agents can reach consensus so that $y_1^{K_i} = \cdots = y_{n_i+q_i}^{K_i} = \varphi(l)/(n_i+q_i)$, where K_i is the number of distinct nonzero eigenvalues of the graph Laplacian matrix $\mathcal{L}_i$ in Cluster i. Finally each agent gets $\varphi(l)$ by multiplying $n_i + q_i$ with $y_j^{K_i}, \forall j = 1, \cdots, n_i + q_i$. For $l = 0$, the above procedures are the same except replacing $\varphi(l)$ with $\varphi(0)$.

Step 3: Projected gradient operation: Each generator agent in Cluster i takes a projected gradient step to minimize its own cost function f_i^j, i.e.,

$$\hat{x}_i^j(l) = P_{\bar{X}_i^j}\left[\varphi(l) - \zeta_l \nabla f_i^j(\varphi(l))\right] \qquad (6.5)$$

where $P_{\bar{X}_i^j}[.]$ is the projection operator onto the local constraint set $\bar{X}_i^j$, ζ_l is a stepsize at iteration l, ∇f_i^j denotes the gradient of the cost function f_i^j.

Step 4: Cluster estimate through average consensus: Through averaging the the estimates of all the agents in Cluster i, a *cluster estimate* $\varphi'_i(l)$ at iteration l is obtained by applying distributed FACA, i.e., $\varphi'_i(l) = \dfrac{\sum\limits_{j=1}^{n_i} \hat{x}_i^j(l)}{n_i}$.

Step 5: Extra manipulation yielding modified cluster estimate: Once each
agent reaches average consensus to the cluster estimate $\varphi'_i(l)$, the leader agent
of the cluster conducts the following simple manipulation to obtain the *modified cluster estimate* $\varphi_i(l)$.

$$\varphi_i(l) = \frac{n_i}{\tilde{n}} \varphi'_i(l) + \frac{(\tilde{n} - n_i)\varphi(l)}{\tilde{n}}, \quad i \in \mathcal{M} \tag{6.6}$$

where $\tilde{n} = \max\{n_1, \cdots, n_m\}$.

Remark 6.2 *In Eqn. (6.6) is an extra manipulation conducted by the leader agent,
which is proposed to achieve convergence of the algorithm with the help of a new idea
on* **virtual agent**, *as analyzed later. Actually, this manipulation is necessary, as some
of our numerical simulation studies show that, without it, the estimated solutions will
not converge to the optimal point if each cluster has a different number of agents.*

Remark 6.3 *Note that a global information $\tilde{n}$ is used in (6.6). But it can be obtained from local information. In fact under Assumption 6.3, the leader agents in
each cluster can easily obtain the number of the agents $n_i, i = 1, \cdots, m$ in other clusters through "network flooding" communication strategy in the graph discovery algorithm proposed in Chapter 2.*

6.3.2 Distributed sequential optimization algorithm

Generally speaking, there are mainly two kinds of sequential communication strategies for a leader agent to send the estimate to its neighboring cluster leader, *namely*,
deterministic [132] and random sequence [133]. In this chapter, we mainly focus on
deterministic communication, specifically, the round-robin communication strategy.
Based on this strategy, in one cycle of iteration each cluster updates the estimate
once in sequence. Therefore, one cycle iteration consists of m clusters of estimate
updating. Without loss of generality, suppose the sequential order is from Cluster 1
to Cluster m in an increasing way.

An illustrative diagram of our proposed distributed optimization algorithm is
shown in Fig. 6.3, where the detailed steps of only Cluster 1 are shown. Now each
corresponding step of Cluster i is described as follows:

Step 0: Initialization: As a starting point, at $l = 0$, the leader agent in Cluster 1
starts estimating the optimal solution by choosing one arbitrary initial value
$\varphi(0) \in \mathbb{R}^n$ and then goes to Step 2.

Step 1: Receiving estimated solution from the leader agent of another cluster: Suppose at iteration step l, for $l \geq 1$, the leader agent in Cluster i, $i \in \mathcal{M}$
receives $\varphi_{i-1}(l)$ sent by the leader agent in Cluster $i - 1$, where $\varphi_{i-1}(l)$ is
called *modified cluster estimate*, as illustrated in Step 5.

Step 2: Information exchange and consensus: Within Cluster i, by letting each
agent exchange information with its neighbors, all the agents reach an average consensus of the received estimate. This process can be summarized

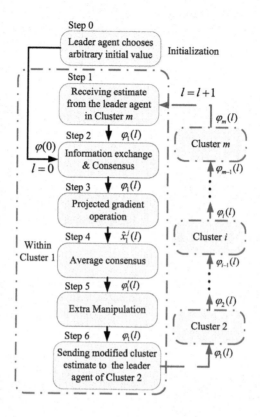

Figure 6.3: A diagram of the proposed distributed sequential optimization

as follows. According to Lemma 2.1, for $l \geq 1$, the initial value can be set as $y_1^0 = \varphi_{i-1}(l), y_2^0 = \cdots = y_{n_i}^0 = 0$. Then by using FACA, after K_i steps, all the agents can reach consensus so that $y_1^{K_i} = \cdots = y_{n_i}^{K_i} = \varphi_{i-1}(l)/n_i$, where K_i is the number of distinct nonzero eigenvalues of the graph Laplacian matrix $\mathcal{L}_i$ in Cluster i. Finally each agent gets $\varphi_{i-1}(l)$ by multiplying n_i with $y_j^{K_i}, \forall j = 1, \cdots, n_i$. For $l = 0$, the above procedures are the same except replacing $\varphi_{i-1}(l)$ with $\varphi(0)$.

Step 3: Projected gradient operation: Each agent in Cluster i takes a projected gradient step to minimize its own cost function f_i^j, i.e.,

$$\hat{x}_i^j(l) = P_{\bar{X}_i^j}\left[\varphi_{i-1}(l) - \zeta_l \nabla f_i^j(\varphi_{i-1}(l))\right] \tag{6.7}$$

where $P_{\bar{X}_i^j}[.]$ is the projection operator onto the set $\bar{X}_i^j$, ζ_l is a stepsize at iteration l, ∇f_i^j denotes the gradient of the cost function f_i^j.

Step 4: Cluster estimate through average consensus: Through averaging the estimates of all the agents in Cluster i, a *cluster estimate* $\varphi'_i(l)$ at iteration l is obtained by applying distributed FACA as follows

$$
\begin{cases}
z_1^{i,j}(l) = w_{jj}(1)\hat{x}_i^j(l) + \sum_{k\in\mathcal{N}_j} w_{jk}(1)\hat{x}_i^k(l) \\
z_2^{i,j}(l) = w_{jj}(2)z_1^{i,j}(l) + \sum_{k\in\mathcal{N}_j} w_{jk}(2)z_1^{i,k}(l) \\
\quad\quad\quad\vdots \\
z_{K_i}^{i,j}(l) = w_{jj}(K_i)z_{K_i-1}^{i,j}(l) + \sum_{k\in\mathcal{N}_j} w_{jk}(K_i)z_{K_i-1}^{i,k}(l) \\
\varphi'_i(l) = z_{K_i}^{i,j}(l)
\end{cases}
\tag{6.8}
$$

where $w_{jj}(s), w_{jk}(s), s = 1,\cdots,K_i$ are the updating gains chosen according to Lemma 2.1. Note that the cluster estimate $\varphi'_i(l)$ is in fact the average of the estimates of local agents in Cluster i, i.e., $\varphi'_i(l) = (\sum_{j=1}^{n_i} \hat{x}_i^j(l))/n_i$.

Step 5: Extra manipulation yielding modified cluster estimate: Once each agent reaches average consensus to the cluster estimate $\varphi'_i(l)$, the leader agent of the cluster conducts the following simple manipulation to obtain the *modified cluster estimate* $\varphi_i(l)$.

$$
\varphi_i(l) = \frac{n_i}{\tilde{n}}\varphi'_i(l) + \frac{(\tilde{n}-n_i)\varphi_{i-1}(l)}{\tilde{n}}, \quad i \in \mathcal{M}
\tag{6.9}
$$

where $\tilde{n} = \max\{n_1,\cdots,n_m\}$.

Step 6: Sending the modified cluster estimate to the leader agent of another cluster: The leader agent sends the modified cluster estimate $\varphi_i(l)$ to the leader agent of Cluster $i+1$.

Let x_l be the estimate after l cycles, then the estimate after another cycle is

$$
x_{l+1} = \varphi_m(l)
\tag{6.10}
$$

where $\varphi_m(l)$ is obtained after sequential update of m clusters according to (6.9) starting with

$$
\varphi_0(l) = x_l
\tag{6.11}
$$

Remark 6.4 *In Assumption 6.2, a general undirected connected communication graph is assumed in each cluster. However, if a cluster, for example Cluster i, has an agent that can directly communicate with all the other agents, this agent can be selected as the leader agent of Cluster i and FACA applied in Steps 2 and 4 can be avoided within this cluster. In this case, the leader agent can directly broadcast its received modified cluster estimate $\varphi_{i-1}(l)$ to all the other agents in Step 2; while in Step 4, Agent j in Cluster i sends its updated estimate $\hat{x}_i^j(l)$, $j = 2,\cdots,n_i$ to the* *leader agent. Then the cluster estimate $\varphi'_i(l)$ can be obtained by the leader agent through averaging all the local estimates, i.e., $\varphi'_i(l) = (\sum_{j=1}^{n_i} \hat{x}_i^j(l))/n_i$.*

Remark 6.5 *Eqn. (6.9) is similar to Eqn. (6.6), which is an extra manipulation proposed to achieve convergence of the algorithm.*

Remark 6.6 *In this section, we only consider deterministic sequential communication strategy. However, our approach can be further developed by considering stochastic Markov chain-based communication strategy proposed in [133].*

6.3.3 Virtual agent

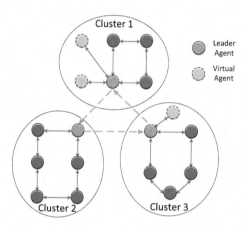

Figure 6.4: Illustration of virtual agent

As mentioned in Remark 6.2 and Remark 6.5 before, extra manipulation (6.6) and (6.9) are conducted by the leader agents in both algorithms, respectively. By further looking into (6.6) or (6.9), it is actually a linear combination of two parts, one is the average consensus value of the n_i agents in Cluster i. The other is the $\tilde{n} - n_i$ copies of previous estimate, which can be interpreted as the effects of certain *virtual agents*. The *virtual agent* proposed for a cluster refers to an imaginary agent that has a constant cost function, i.e., no optimization variable. The main purpose of proposing *virtual agents* is to make the number of agents in each cluster be equal, which is a key idea to establish the convergence of the proposed two algorithms.

Suppose the leader agent in Cluster i adds $\tilde{n} - n_i$ virtual agents and builds up a communication link among them, as shown in Fig. 6.4. By setting the cost function and constraint set of the virtual agents as $f_i^j(x) = C, \bar{X}_i^j = \mathbb{R}^n, \forall i = 1, \cdots, m, j = n_i + 1, \cdots, \tilde{n}$, respectively, where C is some constant, (6.2) can be equivalently reformulated as

$$
\begin{cases}
\min\limits_{x} \sum\limits_{i=1}^{m} \sum\limits_{j=1}^{\tilde{n}} f_i^j(x) \\
s.t. \quad x \in X
\end{cases}
\tag{6.12}
$$

According to Eqn. (6.5) and (6.7), the updates of the virtual agents in Step 3 are shown as follows, respectively

$$
\begin{aligned}
\hat{x}_i^j(l) &= P_{\tilde{X}_i^j}\left[\varphi(l) - \zeta_l \nabla f_i^j(\varphi(l))\right] \\
&= P_{\tilde{X}_i^j}[\varphi(l)] = \varphi(l), \quad j = n_i + 1, \cdots, \tilde{n}.
\end{aligned}
\tag{6.13}
$$

$$
\begin{aligned}
\hat{x}_i^j(l) &= P_{\tilde{X}_i^j}\left[\varphi_{i-1}(l) - \zeta_l \nabla f_i^j(\varphi_{i-1}(l))\right] \\
&= P_{\tilde{X}_i^j}[\varphi_{i-1}(l)] = \varphi_{i-1}(l), \quad j = n_i + 1, \cdots, \tilde{n}.
\end{aligned}
\tag{6.14}
$$

These indicate that the virtual agent does nothing but keeps the modified estimates from previous iterations, which are indeed in the second part in (6.6) and (6.9), respectively.

6.4 Convergence analysis

In this section, we analyze and prove the convergence of the proposed two distributed algorithms.

6.4.1 Distributed synchronous algorithm

With the concept of *virtual agent*, we can augment j in Eqn. (6.4) to $\tilde{n}$, and then we have

$$
\begin{aligned}
\varphi(l) &= \frac{\sum\limits_{i=1}^{m} \varphi_i(l-1)}{m} = \frac{\sum\limits_{i=1}^{m}\left(\frac{n_i}{\tilde{n}}\frac{\sum\limits_{j=1}^{n_1}\hat{x}_i^j(l-1)}{n_i} + \frac{\tilde{n}-n_i}{\tilde{n}}\varphi(l-1)\right)}{m} \\
&= \frac{\sum\limits_{i=1}^{m}\sum\limits_{j=1}^{\tilde{n}}\hat{x}_i^j(l-1)}{\tilde{n}m}
\end{aligned}
\tag{6.15}
$$

Eqn. (6.15) indicates that it is indeed the average of the whole large-scale system with each cluster augmenting its generator agent to the same number $\tilde{n}$.

Theorem 6.1 *Let $\{\hat{x}_i^j(l)\}$ be the estimates generated by the algorithm in (6.4)-(6.6).*

Then the sequences $\{\hat{x}_i^j(l)\}$ converge to the optimal solution $x^\star$ with $x^\star \in \bigcap\limits_{i=1}^{m} \bigcap\limits_{j=1}^{n_i} \bar{X}_i^j$, i.e.,

$$\lim_{l \to \infty} \hat{x}_i^j(l) = \lim_{l \to \infty} x_l = x^\star, \ \forall i \in \mathcal{M}, j \in \mathcal{A}_i$$

if the stepsize ζ_l satisfies that $\zeta_l > 0$, $\sum_l \zeta_l = \infty$ and $\sum_l \zeta_l^2 < \infty$.
Proof: With the help of virtual agent, it is shown that (6.12) is equivalent to (6.2). Then according to Eqn. (6.15), the algorithm in (6.4)-(6.6) is then equivalent to *Theorem 5.1* in Chapter 5. Then the result follows from the proof of *Theorem 5.1*. ■

6.4.2 Distributed sequential algorithm

In this subsection, we prove the convergence of the proposed algorithm with only global constraint, i.e., $X = \bar{X}_i^j = X_g$.

Firstly, we establish a property on the distance between $\hat{x}_i^j(l)$ and a feasible point $z \in X$, i.e., $\left\| \hat{x}_i^j(l) - z \right\|^2$, as stated in Lemma 6.1 below. Next, to facilitate the convergence analysis, (6.16) in Lemma 6.1 is converted to (6.22) by introducing the concept of *virtual agent*, which is the key idea in the convergence analysis. Then, with the help of virtual agent, a property on $\left\| \hat{x}_i^j(l+1) - z \right\|^2$ and $\left\| \hat{x}_i^j(l) - z \right\|^2$ is established in Lemma 6.2. Lastly, the convergence of the proposed algorithm is presented in Theorem 6.2 with proofs.

Lemma 6.1 *Let $\{\hat{x}_i^j(l)\}$ be the sequence generated by (6.7). Then for any $z \in X$ and all $l \geq 0$,*

$$\left\| \hat{x}_i^j(l) - z \right\|^2 \leq \left\| \varphi_{i-1}(l) - z \right\|^2 + \zeta_l^2 \left\| \nabla f_i^j \right\|^2$$
$$- 2\zeta_l \nabla f_i^j (\varphi_{i-1}(l))^T (\varphi_{i-1}(l) - z) - \left\| \phi_i^j(l) \right\|^2 \tag{6.16}$$

where $\xi_i^j(l)$ is the projection error and is defined as

$$\xi_i^j(l) = P_X \left[\varphi_{i-1}(l) - \zeta_l \nabla f_i^j (\varphi_{i-1}(l)) \right]$$
$$- \left(\varphi_{i-1}(l) - \zeta_l \nabla f_i^j (\varphi_{i-1}(l)) \right) \tag{6.17}$$

Proof: For any $z \in X$ and all i, j and $l \geq 0$, we have

$$\left\| \hat{x}_i^j(l) - z \right\|^2 = \left\| P_X \left[\varphi_{i-1}(l) - \zeta_l \nabla f_i^j (\varphi_{i-1}(l)) \right] - z \right\|^2 \tag{6.18}$$

As $z \in X$, by using Lemma 2.4, (6.18) becomes

$$\left\| \hat{x}_i^j(l) - z \right\|^2$$
$$\leq \left\| \varphi_{i-1}(l) - \zeta_l \nabla f_i^j (\varphi_{i-1}(l)) - z \right\|^2 - \left\| \xi_i^j(l) \right\|^2 \tag{6.19}$$

For notation simplicity, we drop $(\varphi_{i-1}(l))$ in $\nabla f_i^j(\varphi_{i-1}(l))$ below. Note that

$$\left\| \varphi_{i-1}(l) - \zeta_l \nabla f_i^j(\varphi_{i-1}(l)) - z \right\|^2$$
$$= \| \varphi_{i-1}(l) - z \|^2 + \zeta_l^2 \left\| \nabla f_i^j \right\|^2 - 2\zeta_l \nabla f_i^{jT} (\varphi_{i-1}(l) - z) \qquad (6.20)$$

Substituting (6.20) into (6.19), the lemma is proved. ■

With the concept of virtual agent, we can augment j in Eqn. (6.7) to $\tilde{n}$, and then Eqn. (6.9) becomes

$$\varphi_i(l) = \varphi'_i(l), \ \ i \in \mathcal{M} \qquad (6.21)$$

Then (6.16) in Lemma 6.1 can be augmented as

$$\left\| \hat{x}_i^j(l) - z \right\|^2$$
$$\leq \| \varphi_{i-1}(l) - z \|^2 + \zeta_l^2 \left\| \nabla f_i^j \right\|^2 - 2\zeta_l \nabla f_i^{jT} (\varphi_{i-1}(l) - z)$$
$$- \left\| \phi_i^j(l) \right\|^2, \ \ i = 1, \cdots, m, j = 1, \cdots, \tilde{n} \qquad (6.22)$$

where j has been augmented to $\tilde{n}$ for all $i \in \mathcal{M}$.

Lemma 6.2 *Let $\{x_l\}$ be the sequence generated by the algorithm in (6.7)-(6.11). Then for any $z \in X$ and all $l \geq 0$,*

$$\tilde{n} \| x_{l+1} - z \|^2 \leq \tilde{n} \| x_l - z \|^2 + \zeta_l^2 \sum_{i=1}^{m} \sum_{j=1}^{\tilde{n}} \left\| \nabla f_i^j \right\|^2$$
$$- 2\zeta_l \sum_{i=1}^{m} \sum_{j=1}^{\tilde{n}} \nabla f_i^j(\varphi_{i-1}(l))^T (\varphi_{i-1}(l) - z) - \sum_{i=1}^{m} \sum_{j=1}^{\tilde{n}} \left\| \xi_i^j(l) \right\|^2. \qquad (6.23)$$

Proof: Summing (6.22) from $i = 1$ to m, $j = 1$ to $\tilde{n}$, we obtain

$$\sum_{i=1}^{m} \sum_{j=1}^{\tilde{n}} \left\| \hat{x}_i^j(l) - z \right\|^2$$
$$\leq \tilde{n} \sum_{i=1}^{m} \| \varphi_{i-1}(l) - z \|^2 + \zeta_l^2 \sum_{i=1}^{m} \sum_{j=1}^{\tilde{n}} \left\| \nabla f_i^j \right\|^2$$
$$- 2\zeta_l \sum_{i=1}^{m} \sum_{j=1}^{\tilde{n}} \nabla f_i^{jT} (\varphi_{i-1}(l) - z) - \sum_{i=1}^{m} \sum_{j=1}^{\tilde{n}} \left\| \xi_i^j(l) \right\|^2 \qquad (6.24)$$

Note that $\|\varphi_0(l) - z\|^2 = \|x_l - z\|^2$, then we have

$$\sum_{i=1}^{m}\sum_{j=1}^{\tilde{n}}\left\|\hat{x}_i^j(l) - z\right\|^2$$

$$\leq \tilde{n}\|x_l - z\|^2 + \tilde{n}\sum_{i=2}^{m}\|\varphi_{i-1}(l) - z\|^2 + \zeta_l^2\sum_{i=1}^{m}\sum_{j=1}^{\tilde{n}}\left\|\nabla f_i^j\right\|^2$$

$$- 2\zeta_l\sum_{i=1}^{m}\sum_{j=1}^{\tilde{n}}\nabla f_i^{j^T}(\varphi_{i-1}(l) - z) - \sum_{i=1}^{m}\sum_{j=1}^{\tilde{n}}\left\|\xi_i^j(l)\right\|^2 \qquad (6.25)$$

According to (6.8) and (6.21), we have

$$\tilde{n}\sum_{i=2}^{m}\|\varphi_{i-1}(l) - z\|^2 = \tilde{n}\sum_{i=2}^{m}\|\varphi'_{i-1}(l) - z\|^2$$

$$= \tilde{n}\sum_{i=2}^{m}\left\|\frac{\sum_{j=1}^{\tilde{n}}(\hat{x}_{i-1}^j(l) - z)}{\tilde{n}}\right\|^2 \leq \sum_{i=2}^{m}\sum_{j=1}^{\tilde{n}}\left\|\hat{x}_{i-1}^j(l) - z\right\|^2 \qquad (6.26)$$

where in above inequality we use the relation

$$\left\|\frac{a_1 + \cdots + a_n}{n}\right\|^2 \leq \frac{\|a_1\|^2 + \cdots + \|a_n\|^2}{n} \qquad (6.27)$$

Substituting (6.26) into (6.25), we have

$$\sum_{i=1}^{m}\sum_{j=1}^{\tilde{n}}\left\|\hat{x}_i^j(l) - z\right\|^2$$

$$\leq \tilde{n}\|x_l - z\|^2 + \sum_{i=2}^{m}\sum_{j=1}^{\tilde{n}}\left\|\hat{x}_{i-1}^j(l) - z\right\|^2 + \zeta_l^2\sum_{i=1}^{m}\sum_{j=1}^{\tilde{n}}\left\|\nabla f_i^j\right\|^2$$

$$- 2\zeta_l\sum_{i=1}^{m}\sum_{j=1}^{\tilde{n}}\nabla f_i^{j^T}(\varphi'_{i-1}(l) - z) - \sum_{i=1}^{m}\sum_{j=1}^{\tilde{n}}\left\|\xi_i^j(l)\right\|^2 \qquad (6.28)$$

By applying the relation (6.27) to the first term on the left-hand side of (6.28), we obtain

$$\sum_{j=1}^{\tilde{n}}\left\|\hat{x}_m^j(l) - z\right\|^2 \geq \tilde{n}\left\|\varphi'_m(l) - z\right\|^2 = \tilde{n}\|x_{l+1} - z\|^2 \qquad (6.29)$$

Combining (6.28) and (6.29), the lemma is proved. ■

Remark 6.7 *It is worthy to point out that the term* $\sum_{i=2}^{m} \sum_{j=1}^{\tilde{n}} \left\| \hat{x}_{i-1}^{j}(l) - z \right\|^2$ *on the right-hand side of (6.28) cannot be canceled by the term on the left if (6.9) is not conducted by the leader agent, i.e., the virtual agents are not introduced in each cluster to make the number of agents be equal to* $\tilde{n}$. *In other words, the extra manipulation (6.9) makes the two terms equal and thus cancel each other so that Lemma 6.2 can be successfully established. This is one of the key ideas to establish the convergence of the proposed algorithm below.*

Theorem 6.2 *Let* $\{x_l\}$ *be the estimates generated by the algorithm in (6.7)-(6.11). Then the sequences* $\{x_l\}$ *and* $\{\hat{x}_i^j(l)\}$ *converge to the optimal solution* $x^\star$ *with* $x^\star \in X$, *i.e.,*

$$\lim_{l \to \infty} \hat{x}_i^j(l) = \lim_{l \to \infty} x_l = x^\star, \ \forall i \in \mathcal{M}, j \in \mathcal{A}_i$$

if the stepsize ζ_l *satisfies that* $\zeta_l > 0$, $\sum_l \zeta_l = \infty$ *and* $\sum_l \zeta_l^2 < \infty$.

Proof: In this proof, we first prove that $f(x_l)$ converges to $f(x^\star)$ as $l \to \infty$, where $f(x_l) = \sum_{i=1}^{m} \sum_{j=1}^{\tilde{n}} f_i^j(x_l)$. We then show the sequences $\{x_l\}$ and $\{\hat{x}_i^j(l)\}$ converge to the same optimal point $x^\star$.

According to the well-known gradient inequality, see for example [131], we have

$$-\nabla f_i^j \left(\varphi_{i-1}(l) \right)^T \left(\varphi_{i-1}(l) - z \right) \leq f_i^j(z) - f_i^j \left(\varphi_{i-1}(l) \right) \tag{6.30}$$

Substituting (6.30) into (6.23) in Lemma 6.2, we have

$$\tilde{n} \|x_{l+1} - z\|^2 \leq \tilde{n} \|x_l - z\|^2 + \zeta_l^2 \sum_{i=1}^{m} \sum_{j=1}^{\tilde{n}} \left\| \nabla f_i^j \right\|^2$$

$$+ 2\zeta_l \sum_{i=1}^{m} \sum_{j=1}^{\tilde{n}} \left(f_i^j(z) - f_i^j \left(\varphi_{i-1}(l) \right) \right) - \sum_{i=1}^{m} \sum_{j=1}^{\tilde{n}} \left\| \xi_i^j(l) \right\|^2 \tag{6.31}$$

Adding and subtracting $f_i^j \left(\varphi_0(l) \right)$ in the third term in the right-hand side of (6.31) yields

$$\sum_{i=1}^{m} \sum_{j=1}^{\tilde{n}} \left(f_i^j(z) - f_i^j \left(\varphi_{i-1}(l) \right) \right)$$

$$= \sum_{i=1}^{m} \sum_{j=1}^{\tilde{n}} \left(f_i^j(z) - f_i^j \left(\varphi_0(l) \right) + f_i^j \left(\varphi_0(l) \right) - f_i^j \left(\varphi_{i-1}(l) \right) \right)$$

$$= f(z) - f(x_l) + \sum_{i=2}^{m} \sum_{j=1}^{\tilde{n}} \left(f_i^j \left(\varphi_0(l) \right) - f_i^j \left(\varphi_{i-1}(l) \right) \right)$$

$$\leq f(z) - f(x_l) + \sum_{i=2}^{m} \sum_{j=1}^{\tilde{n}} \left\| \nabla f_i^j \left(\varphi_0(l) \right) \right\| \left\| \varphi_0(l) - \varphi_{i-1}(l) \right\|. \tag{6.32}$$

For $\|\varphi_0 - \varphi_{i-1}\|$, where l is dropped for notation simplicity, it can be shown that

$$\|\varphi_{i-1} - \varphi_0\| = \|\varphi'_{i-1} - \varphi_0\| = \left\| \frac{\sum\limits_{j=1}^{\tilde{n}} \left(\hat{x}^j_{i-1} - \varphi_0 \right)}{\tilde{n}} \right\|$$

$$\leq \frac{1}{\tilde{n}} \sum_{j=1}^{\tilde{n}} \left\| \hat{x}^j_{i-1} - \varphi_0 \right\|$$

$$= \frac{1}{\tilde{n}} \sum_{j=1}^{\tilde{n}} \left\| \xi^j_{i-1} + \varphi_{i-2} - \zeta_l \nabla f^j_{i-1} - \varphi_0 \right\|$$

$$\leq \frac{1}{\tilde{n}} \sum_{j=1}^{\tilde{n}} \left(\left\| \xi^j_{i-1} \right\| + \zeta_l \left\| \nabla f^j_{i-1} \right\| + \| \varphi_{i-2} - \varphi_0 \| \right)$$

$$\leq \cdots$$

$$\leq \frac{1}{\tilde{n}} \sum_{r=1}^{i-1} \sum_{j=1}^{\tilde{n}} \left(\left\| \xi^j_r \right\| + \zeta_l \left\| \nabla f^j_r \right\| \right) \tag{6.33}$$

Substituting (6.33) into (6.32), and using Assumption 6.1, we have

$$\sum_{i=1}^{m} \sum_{j=1}^{\tilde{n}} \left(f^j_i(z) - f^j_i \left(\varphi_{i-1}(l) \right) \right)$$

$$\leq \sum_{i=2}^{m} \sum_{j=1}^{\tilde{n}} \left\| \nabla f^j_i(\varphi_0(l)) \right\| \left(\frac{1}{\tilde{n}} \sum_{r=1}^{i-1} \sum_{j=1}^{\tilde{n}} \left(\left\| \xi^j_r \right\| + \zeta_l \left\| \nabla f^j_r \right\| \right) \right)$$

$$+ f(z) - f(x_l)$$

$$\leq \sum_{i=2}^{m} \sum_{j=1}^{\tilde{n}} L \left(\frac{1}{\tilde{n}} \sum_{r=1}^{i-1} \sum_{j=1}^{\tilde{n}} \left(\left\| \xi^j_r \right\| + \zeta_l L \right) \right) + f(z) - f(x_l)$$

$$\leq \sum_{i=1}^{m} \sum_{j=1}^{\tilde{n}} (m-i)L \left\| \xi^j_i \right\| + \frac{\tilde{n}m(m-1)L^2}{2} \zeta_l + f(z) - f(x_l). \tag{6.34}$$

Substituting (6.34) into (6.31), we obtain

$$\|x_{l+1} - z\|^2 \leq \|x_l - z\|^2 + \zeta_l^2 C + \frac{2}{\tilde{n}} \zeta_l \left[f(z) - f(x_l) \right]$$

$$+ \frac{1}{\tilde{n}} \sum_{i=1}^{m} \sum_{j=1}^{\tilde{n}} \left(2(m-i)L\zeta_l \left\| \xi^j_i \right\| - \left\| \xi^j_i \right\|^2 \right) \tag{6.35}$$

where $C = m^2 L^2$.

Note that

$$2(m-i)L\zeta_l \left\| \xi_i^j \right\| - \left\| \xi_i^j \right\|^2$$
$$= -\left(\left\| \xi_i^j \right\| - (m-i)L\zeta_l \right)^2 + (m-i)^2 L^2 \zeta_l^2$$
$$\leq (m-i)^2 L^2 \zeta_l^2, \quad i = 1, \cdots, m-1 \tag{6.36}$$

Combining (6.35) and (6.36), we have

$$\|x_{l+1} - z\|^2 \leq \|x_l - z\|^2 + \zeta_l^2 C + \frac{2}{\tilde{n}} \zeta_l [f(z) - f(x_l)]$$
$$+ \sum_{i=1}^{m-1} (m-i)^2 L^2 \zeta_l^2 - \sum_{j=1}^{\tilde{n}} \left\| \xi_m^j \right\|^2$$
$$\leq \|x_l - z\|^2 + \zeta_l^2 D + \frac{2}{\tilde{n}} \zeta_l [f(z) - f(x_l)] - \sum_{j=1}^{\tilde{n}} \left\| \xi_m^j \right\|^2 \tag{6.37}$$

where $D = C + \sum_{i=1}^{m-1} (m-i)^2 L^2$.

Summing the preceding relations for l from G to H for some arbitrary G and H with $G < H$, and setting $z = x^\star$ we have

$$\|x_{H+1} - x^\star\|^2 + \sum_{l=G}^{H} \sum_{j=1}^{\tilde{n}} \left\| \xi_m^j(l) \right\|^2$$
$$+ \frac{2}{\tilde{n}} \sum_{l=G}^{H} \zeta_l [f(x_l) - f(x^\star)] \leq \|x_G - x^\star\|^2 + D \sum_{l=G}^{H} \zeta_l^2 \tag{6.38}$$

By setting $G = 0$ and letting $H \to \infty$, (6.38) becomes

$$\|x_\infty - x^\star\|^2 + \frac{2}{\tilde{n}} \sum_{l=0}^{\infty} \zeta_l [f(x_l) - f(x^\star)] + \sum_{l=0}^{\infty} \sum_{j=1}^{\tilde{n}} \left\| \xi_m^j(l) \right\|^2$$
$$\leq \|x_0 - x^\star\|^2 + D \sum_{l=0}^{\infty} \zeta_l^2 \tag{6.39}$$

As $\sum_l \zeta_l^2 < \infty$, then the right-hand side of (6.39) is bounded. Hence

$$\frac{2}{\tilde{n}} \sum_{l=0}^{\infty} \zeta_l [f(x_l) - f(x^\star)] + \sum_{l=0}^{\infty} \sum_{j=1}^{\tilde{n}} \left\| \xi_m^j(l) \right\|^2 \leq \infty$$

which also implies that

$$\frac{2}{\tilde{n}} \sum_{l=0}^{\infty} \zeta_l \left[f(x_l) - f(x^\star) \right] \leq \infty \tag{6.40}$$

$$\sum_{l=0}^{\infty} \sum_{j=1}^{\tilde{n}} \left\| \xi_m^j(l) \right\|^2 \leq \infty \tag{6.41}$$

Note that (6.7)-(6.10) indicates that x_l is a linear combination of $\hat{x}_i^j(l)$, and $\hat{x}_i^j(l) \in X$ for all $l \geq 0$. Also X is a convex set, thus we conclude that $x_l \in X$, $l \geq 0$, which implies that $f(x_l) - f(x^\star) \geq 0$ for all $l \geq 0$. Note that $\sum_l \zeta_l = \infty$. Then from (6.40) we obtain

$$\lim_{l \to \infty} f(x_l) - f(x^\star) = 0$$

which yields

$$\liminf_{l \to \infty} f(x_l) = f(x^\star) \tag{6.42}$$

Next we show that the sequence $\{x_l\} = \{\varphi_0(l)\}$ converges to the same optimal point $x^\star$.

Motivated by the convergence analysis in [130], by taking limsup as $H \to \infty$ in (6.38) and then liminf as $G \to \infty$, we obtain for any $z^\star \in X^\star$

$$\limsup_{H \to \infty} \|x_{H+1} - z^\star\|^2 \leq \liminf_{G \to \infty} \|x_G - z^\star\|^2$$

which implies that the scalar sequence $\{\|x_l - z^\star\|\}$ is convergent for every $z^\star \in X^\star$. As $\liminf_{l \to \infty} f(x_l) = f(x^\star)$, it means that the limit point of $\{x_l\}$ must belong to $X^\star$, which is denoted as $x^\star$. Since $\{\|x_l - z^\star\|\}$ is convergent for $z^\star = x^\star$, it follows that $\lim_{l \to \infty} x_l = x^\star$. This means the sequence $\{x_l\} = \{\varphi_0(l)\}$ converges to the optimal solution $x^\star$.

Lastly we show that the sequence $\{\hat{x}_i^j(l)\}, \forall i \in \mathcal{M}, j \in \mathcal{A}_i$ also converges to the optimal point $x^\star$.

Note that $\liminf_{l \to \infty} \zeta_l = 0$. From Eqn. (6.7), we have

$$\lim_{l \to \infty} \hat{x}_i^j(l) = \lim_{l \to \infty} P_X[\varphi_{i-1}(l)] \tag{6.43}$$

As $\lim_{l \to \infty} \varphi_0(l) = \lim_{l \to \infty} x_l = x^\star \in X$, thus we obtain $\lim_{l \to \infty} \varphi_i \in X$, $i = 1, \cdots, m-1$. Then (6.43) becomes

$$\lim_{l \to \infty} \hat{x}_i^j(l) = \lim_{l \to \infty} \varphi_{i-1}(l) \tag{6.44}$$

According to (6.7), we have

$$\lim_{l \to \infty} \varphi_i(l) = \lim_{l \to \infty} \varphi_{i-1}(l) \tag{6.45}$$

which implies that $\lim_{l \to \infty} \hat{x}_i^j(l) = \lim_{l \to \infty} \varphi_i(l) = \lim_{l \to \infty} x_l = x^\star, \forall i \in \mathcal{M}, j \in \mathcal{A}_i$. This means the sequence $\{\hat{x}_i^j(l)\}, \forall i \in \mathcal{M}, j \in \mathcal{A}_i$ also converges to the optimal point $x^\star$. This completes the proof. ■

6.5 Economic Dispatch in Multi-Area Power System

Economic dispatch (ED) in a multi-area power system is one of the key problems in power system research. Conventionally, such problem is solved in a centralized way. Although there have been some distributed ED optimization methods, e.g., [85]-[91], most of them focus on a single-area system so that synchronous communication among generators in the area is possible. However, for an interconnected multi-area power system in which different areas locate distantly, synchronous communication may not be proper. In this section, we apply the proposed two distributed optimization methods to solve the ED problem in a multi-area power system.

6.5.1 Problem statement

Mathematically speaking, the objective of an ED problem is to minimize the total generation cost subject to the demand supply constraint as well as the generator constraints [167]. Different from the work in Chapter 5, which mainly deals with the ED problem with a single area, we consider a multi-area power system consisting of both thermal generators (TGs) and loads in this chapter. Thus the main goal of ED is to minimize the system cost given by

$$C = \sum_{i \in \mathcal{M}} \sum_{j \in S_{G_i}} g_i^j(P_i^j) \tag{6.46}$$

where $\mathcal{M}$ is the set of the areas, S_{G_i} is the set of TGs in the ith area, P_i^j is the scheduled power output of the jth TG, $j \in S_{G_i}$ in the ith area.

The cost of conventional TG is usually approximated by a quadratic function [167]:

$$g_i^j(P_i^j) = \alpha_i^j P_i^{j^2} + \beta_i^j P_i^j + \gamma_i^j \tag{6.47}$$

where α_i^j, β_i^j and γ_i^j are the cost coefficients of the jth TG in the ith area.

Then by ignoring the transmission losses and constraints, and considering both generator constraints and demand supply constraint, the multi-area ED problem can be formulated as

$$\begin{cases} \min_{P_i^j} \sum_{i \in \mathcal{M}} \sum_{j \in S_{G_i}} g_i^j(P_i^j) \\ s.t. \sum_{i \in \mathcal{A}} \sum_{j \in S_{G_i}} P_i^j = P_d \\ P_i^{j^{\min}} \leq P_i^j \leq P_i^{j^{\max}}, j \in S_{G_i}, \forall i \in \mathcal{M} \end{cases} \tag{6.48}$$

where $P_i^{j\,\text{min}}$ and $P_i^{j\,\text{max}}$ are the lower and upper bounds of the jth TG in the ith area, P_d is the total load demand satisfying $\sum_{i\in\mathcal{M}}\sum_{j\in S_{G_i}} P_i^{j\,\text{min}} \le P_d \le \sum_{i\in\mathcal{A}}\sum_{k\in S_{G_i}} P_i^{j\,\text{max}}$.

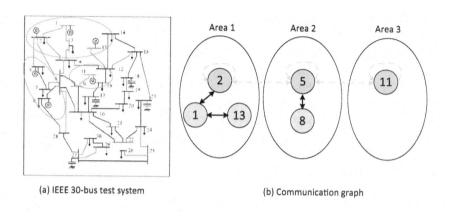

(a) IEEE 30-bus test system   (b) Communication graph

Figure 6.5: IEEE 30-bus test system

Suppose there are $m = |\mathcal{M}|$ areas in the power system, consisting of $n_i = |S_{G_i}|$ TGs in the ith area. We treat every generator as an "agent," and each agent is assigned a unique ID. Denote the generated power in area i in a global vector as $x_i = \begin{bmatrix} P_i^1 & \cdots & P_i^j & \cdots & P_i^{n_i} \end{bmatrix}^T$. Taking the whole system into account, the global scheduled generated power for an A-area power system can be written as $x = \begin{bmatrix} x_1 & x_2 & \cdots & x_A \end{bmatrix}^T \in \mathbb{R}^n$, where $n = \sum_{i=1}^{m} n_i$ is the total number of generators. Then the cost function f_i^j and constraint set $\bar{X}_i^j$ of agent j, $j \in S_{G_i}$ in i^{th} area are denoted as follows, respectively.

$$f_i^j(x) = g_i^j([x]_h), \quad \forall i = 1,\cdots,m, \; j = 1,\cdots,n_i \tag{6.49}$$

$$\bar{X}_i^j = X_i^j \cap X_g, \quad \forall i = 1,\cdots,m, \; j = 1,\cdots,n_i \tag{6.50}$$

where $h = \sum_{p=1}^{i-1} n_p + j$, $[x]_h$ denotes the hth element of vector x. The global constraint set

$$X_g = \{x | \mathbf{1}^T x = P_d\} \tag{6.51}$$

is the demand supply constraint known to all the agents, where $\mathbf{1} \in \mathbb{R}^n$ is a $n \times 1$ vector with all the elements being 1. The local constraint set is

$$X_i^j = \{x | P_i^{j\,\text{min}} \le [x]_h \le P_i^{j\,\text{max}}\}. \tag{6.52}$$

Both the local cost function $f_i^j(x)$ and constraint set $\bar{X}_i^j$ are only known to Agent j in Cluster i and cannot be shared with other agents, which in reality is the confidential information of the local TGs.

Then (6.48) can be reformulated as the general form (6.1).

6.5.2 Case studies

In this subsection, the IEEE 30-bus system is chosen as a test system. The generator and load bus parameters are adopted from [90], which are also listed in Table 5.5 and Table 5.6, respectively. First, the 6 generators are divided into 3 clusters according to the distance of their physical connections, i.e., generators 1, 2 and 13 are in Cluster 1; generators 5 and 8 are in Cluster 2, and generator 11 is in Cluster 3, shown in Fig. 6.5. Without loss of generality, generators 2, 5, and 11 are assigned as the respective leader agents. The communication graph between each leader agent is shown in Fig. 6.5 (b). Initially it is easy to obtain from Table 5.6 that the total load demand $P_d = \sum_{i=1}^{30} P_i^d = 283.4MW$. Several case studies are presented and discussed here. In the first 2 case studies, the round-robin communication strategy is applied, where the situations without and with generator constraints are investigated. Lastly, we apply the MC-based random communication strategy to our algorithm and compare the performance to the previous round-robin one.

6.5.2.1 Case study 1: Distributed synchronous algorithm

In this case study, we apply the distributed synchronous algorithm to solve the multi-area ED problem formulated in (6.48). The initial value x^0 is chosen as $x^0 = \begin{bmatrix} 0 & \cdots & 0 \end{bmatrix}^T \in \mathbb{R}^6$. To meet the stepsize condition, a diminishing stepsize is chosen as $\zeta_l = \frac{\zeta_0}{l}, l \geq 1$, where ζ_0 is the initial stepsize and is set as $\zeta_0 = 700$ in this example. The simulation results are shown in Fig. 6.6. From Fig. 6.6 (a), we see that the estimates of all the agents $\hat{x}_i^j$ reach consensus and also converge to the optimal point $x^\star = \begin{bmatrix} 185.10 & 46.95 & 19.14 & 10.06 & 10.06 & 12.06 \end{bmatrix}^T$, which indicates the optimal power allocation for each generator is that in area 1 $P_{G_1} = 185.10MW$, $P_{G_2} = 46.95MW$, $P_{G_{13}} = 12.06MW$, in area 2 $P_{G_5} = 19.14MW$, $P_{G_8} = 10.06MW$, in area 3 $P_{G_{11}} = 10.06MW$ with the total cost $C = 767.6021\$$. To clearly demonstrate the result, the estimate of TG 1 is detailed in Fig. 6.6 (b).

6.5.2.2 Case study 2: Distributed sequential algorithm

In this case study, we apply the distributed sequential algorithm to solve the multi-area ED problem formulated in (6.48). The initial value and initial step size are chosen as the same as those in Case Study 1. The simulation results are shown in Fig. 6.7. The iteration process of estimating is shown in Fig. 6.7, which converges to the same optimal values after $k = 3 \times 10^4$ iterations, the same as those in Case study 1.

These two cases show that our proposed two distributed algorithms can both handle the multi-area ED problem.

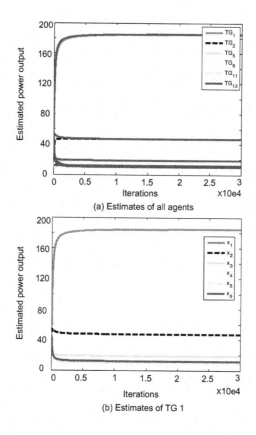

(a) Estimates of all agents

(b) Estimates of TG 1

Figure 6.6: Simulation results with distributed synchronous algorithm

6.5.2.3 Case study 3: Distributed sequential algorithm with random communication strategy

In this case study, we evaluate the performances of distributed sequential algorithm with random communication strategy. One typical random communication strategy is based on the Markov chain (MC) jumper rule [133]. Suppose currently Cluster i is conducting the optimization, the leader agent in Cluster i sends the estimate to one of its neighboring clusters (including itself) according to transmission probability. Referring to [174], the element of the transmission probability matrix of the MC $\mathcal{P}_{ik}, \forall i, k \in \mathcal{M}$ is set as

$$
\mathcal{P}_{ik} = \begin{cases} \min\left\{\frac{1}{a_i}, \frac{1}{a_k}\right\}, & k \in \mathcal{N}_i \\ 1 - \sum_{k \in \mathcal{N}_i} \min\left\{\frac{1}{a_i}, \frac{1}{a_k}\right\}, & k = i \\ 0, & otherwise \end{cases} \tag{6.53}
$$

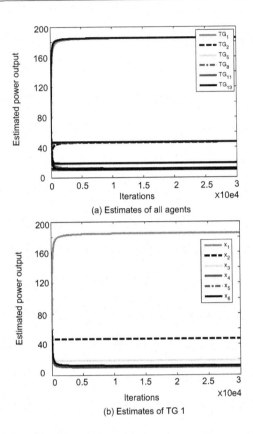

(a) Estimates of all agents

(b) Estimates of TG 1

Figure 6.7: Simulation results with distributed sequential algorithm

where a_i is the number of the neighboring leader agents of the leader agent in cluster i. It is easy to prove that the transmission probability matrix $\mathcal{P}_{ik}$ is a doubly stochastic matrix, which means that all the groups are "visited" often equally in the probability point of view.

According to (6.53) and the communication graph in Fig. 6.5 (b), the MC transmission probability matrix $\mathcal{P}$ can be easily obtained as

$$\mathcal{P} = \begin{bmatrix} 1/2 & 1/2 & 0 \\ 1/2 & 0 & 1/2 \\ 0 & 1/2 & 1/2 \end{bmatrix}$$

Just taking the leader agent in area 1 (generator 2) as an example, the obtained matrix $\mathcal{P}$ indicates that it has equal probabilities ($\mathcal{P}_{11} = \mathcal{P}_{12} = 1/2$) to send the group estimate to generator 5 in area 2 or keep it to itself in the next iteration step. In the simulation, a random variable in range [0 1] is defined and produced randomly by computer. At one particular iteration, if it is greater than $1/2$, then send the estimate to generator 5 in area 2, otherwise keep it to itself.

The simulation results of MC-based random communication strategy are shown in Fig. 6.8. Comparing these two strategies, both can ensure convergence to the optimal value almost with the same convergence speed. However, the estimated trajectory of the random communication is not as "smooth" as the round-robin one.

6.5.2.4   Case study 4: Fast gradient

Note that, in step 3 of the distributed sequential algorithm, only the current iteration information is used. Motivated by the accelerated gradient method in [175, 176], we hope to speed up the convergence by adding a momentum term $\rho_i^j(l) = \hat{x}_i^j(l) - \hat{x}_i^j(l-1)$ in (6.7), i.e.,

$$\hat{x}_i^j(l) = P_{\hat{X}_i^j}\left[\varphi_{i-1}(l) - \zeta_l \nabla f_i^j(\varphi_{i-1}(l)) + \eta \rho_i^j(l)\right] \tag{6.54}$$

where η is called the speed-up gain and usually chosen as $0 < \eta < 1$. The algorithm with (6.54) replacing (6.7) is tested by simulation studies by setting $\eta = 0.4$ in this case study. Besides, to have a more distinguished comparison, we have the same initial conditions except that we set a small initial stepsize as $\zeta_0 = 100$. The simulation results of the estimates of agent1 in cluster 1 are shown in Fig. 6.9, where the dashed line represents the result with the modification in (6.54). Compared to the original one (solid line), it is observed that x_1 converges to its optimal value 185.1 much faster and thus we also name the modified algorithm the fast gradient method. However, establishing a convergence property for this algorithm is challenging and thus still remains an open problem. Therefore it deserves further research.

Acknowledgment

Reprinted from *Automatica*, vol. 77, Fanghong Guo, Changyun Wen, Jianfeng Mao, Guoqi Li, and Yong-Duan Song,"A Distributed Hierarchical Algorithm for Multi-Cluster Constrained Optimization," pp.230 - 238, Copyright (2017), with permission from Elsevier.

©2017 IEEE. Required, with permission, from Fanghong Guo, Changyun Wen, Lantao Xing, "A Distributed Algorithm for Economic Dispatch in a Large-scale Power," 2016 14th International Conference on Control, Automation, Robotics and Vision (ICARCV), Phuket, Thailand, 2016.

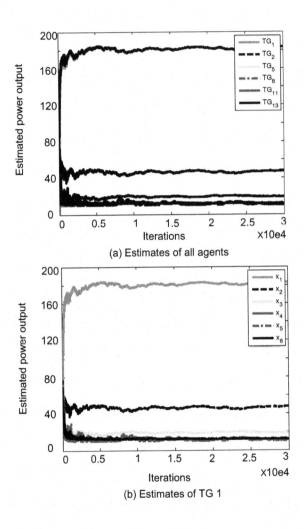

(a) Estimates of all agents

(b) Estimates of TG 1

Figure 6.8: Simulation results of distributed sequential algorithm with random communication strategy

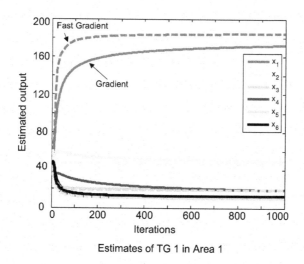

Estimates of TG 1 in Area 1

Figure 6.9: Simulation results of distributed sequential algorithm with fast gradient method

Chapter 7

Hierarchical Decentralized Architecture for ED Problem

In Chapter 5 and Chapter 6, the distributed algorithms for both single-area and multi-area ED are proposed. In this chapter, a different hierarchical decentralized architecture is proposed to solve this problem. An extra coordinator agent is employed to coordinate all the local generator agents. Also, it also takes responsibility to handle the global demand supply constraint based on a concept called *dummy agent*. In this way, different from existing distributed algorithms, the global demand supply constraint and local generation constraints are handled separately, which would greatly reduce the computational complexity. It is theoretically shown that under proposed hierarchical decentralized optimization architecture, each local generator agent can obtain the optimal solution in a decentralized fashion. Several case studies implemented on the IEEE 30-bus and the IEEE 118-bus are discussed and tested to validate the proposed method.

7.1   Introduction

Recently, distributed algorithms have been widely studied and developed to solve the ED problem. In contrast to centralized optimization methods, distributed optimization decomposes the central problem into several local optimization problems. In each local problem, the agent only has access to its own cost function and constraint set. With proper communication with its neighboring agents, each agent can cooperatively solve the ED problem in a distributed manner. Among existing distributed ED algorithms, incremental cost or gradient-based methods are most popular due to their simplicity and ease of implementation. It is analyzed that with only global constraint, each agent will obtain the optimal solution if their incremental costs reach consensus [85]. Based on this idea, several distributed incremental cost consensus algorithms are proposed; see [85]-[87]. Apart from incremental cost consensus algorithms, it is reported that distributed gradient-based methods have also been applied to solve ED problem [89]-[90]. However, these methods require that the global equality constraint should be strictly met when choosing the initial values. To avoid this, an initial-free projected gradient based ED method is proposed in Chapter 5 of this book, where an internal finite-time average consensus algorithm (FACA) is applied in each iteration. Note that both the step number of FACA and the dimension of communication information depend on the size of the whole system, specifically, the number of agents. When the whole system becomes larger and larger, the algorithm in Chapter 5 may not be proper, where there would be a disaster not only in the communication steps for FACA, but also the communication information dimension. In addition, in Chapter 5, the global constraints are handled in each local agent, which will lead to a heavy computational cost when the system becomes large.

In order to overcome these disadvantages, in this chapter, a new hierarchical decentralized optimization architecture is proposed for the ED problem. Similar to existing distributed algorithms, we treat each generator as an agent. In addition, an extra coordinator agent is employed on top of each agent, which plays roles mainly in two aspects, i.e., coordination among all the agents and dealing with the global constraints. In order to reduce the communication links, we propose to divide all the agents into clusters. Each cluster is assigned one leader agent, which is assumed to be able to communicate with all the other agents in the same cluster. Then a hierarchical decentralized optimization scheme is proposed to solve the ED problem. The local agent updates its scheduled power output only based on its own cost function and constraint set. Then it sends its update to the leader agent. Afterwards, the leader agent forms all the local estimates as a vector to obtain a cluster estimate and then sends it to the coordinator agent. The coordinator agent returns the cluster estimate to each leader agent after doing a coordination manipulation on the overall estimates. It is theoretically shown that with a proposed hierarchical optimization scheme, each local generator agent can obtain the optimal solution iteratively.

It is worth pointing out that there are mainly two ways to deal with the global constraint in existing distributed algorithms. For distributed algorithms in [89], [90], the global constraint is guaranteed by choosing stringent initial values. While for the method in Chapter 5, this is handled in each local agent by projecting the local estimate into the intersection set of the global constraint set and local constraint set. In this paper, we handle the global constraint set in the coordinator agent by proposing a new concept named *dummy agent*. In contrast to Chapter 5, this approach enables us to deal with the global constraint (demand supply constraint) and local constraints (generator constraint) separately. This would greatly reduce the computational complexity in each iteration, as the projection operation of the intersection of several constraint sets are usually complicated. Also, different from [89], [90], the initial values in our proposed algorithm can be chosen arbitrarily. Furthermore, there is also no need for the estimates in every iteration to be within the feasible set as those in [85]-[87], as long as the proposed algorithm ensures that the estimates of all agents converge to the optimal values and also meet the constraints asymptotically. Such an idea results in a simpler algorithm that is easier to implement.

The advantages of the proposed algorithm can be summarized as follows: 1) Compared to the previous work in Chapter 5, the internal FACA communication is avoided, thus the number of communications and communication links can be reduced. Furthermore, the dimension of communication information for each local agent decreases from n (the total number of agents) to 1; 2) Different from the algorithm in Chapter 5, the global constraint is only known and handled by the coordinator agent, which makes the algorithm much simpler; 3) In contrast to the distributed gradient methods [89]-[90], where the stringent equality constraint should be met when choosing the initial values, our proposed algorithm allows the initial values to be arbitrary.

The remainder of this chapter is organized as follows. In Section 7.2, the ED problem is formulated as a multi-cluster optimization problem. The main results of proposed hierarchical decentralized ED strategy is presented in Section 7.3. Several case studies on the IEEE 30-bus and IEEE 118-bus power system illustrate the effectiveness of the proposed method in Section 7.4.

7.2   Problem Formulation

Considering a large-scale power system consisting of both thermal generators (TGs) and loads, the main goal of ED is to minimize the generation cost. We assume that the TGs are divided into several clusters, then the ED problem can be formulated as

$$C = \sum_{i \in \mathcal{M}} \sum_{j \in S_{Gi}} f_i^j(P_i^j) \tag{7.1}$$

where $\mathcal{M}$ is the set of clusters, S_{Gi} is the set of TGs in the ith cluster, f_i^j and P_i^j are the cost function and the scheduled power output of the jth TG in Cluster i, respectively.

The cost of TG is usually approximated by a quadratic function [85]:

$$f_i^j(P_i^j) = \alpha_i^j P_i^{j^2} + \beta_i^j P_i^j + \gamma_i^j \tag{7.2}$$

where α_i^j, β_i^j, and γ_i^j are the cost coefficients of the jth TG in the Cluster i.

Then by ignoring the transmission losses and constraints, and considering both generator constraints and demand supply constraint, the multi-cluster ED problem can be formulated as

$$
\begin{cases}
\min\limits_{P_i^j} \sum\limits_{i \in \mathcal{M}} \sum\limits_{j \in S_{G_i}} f_i^j(P_i^j) \\
s.t. \sum\limits_{i \in \mathcal{M}} \sum\limits_{j \in S_{G_i}} P_i^j = P_d \\
P_i^{j\,\mathrm{min}} \leq P_i^j \leq P_i^{j\,\mathrm{max}}, j \in S_{G_i}, i \in \mathcal{M}
\end{cases}
\tag{7.3}
$$

where $P_i^{j\,\mathrm{min}}$ and $P_i^{j\,\mathrm{max}}$ are the lower and upper bounds of the jth TG in the ith cluster, respectively, P_d is the total load demand satisfying $\sum\limits_{i \in \mathcal{M}} \sum\limits_{j \in S_{G_i}} P_i^{j\,\mathrm{min}} \leq P_d \leq$ $\sum\limits_{i \in \mathcal{M}} \sum\limits_{j \in S_{G_i}} P_i^{j\,\mathrm{max}}$.

Suppose there are $m = |\mathcal{M}|$ clusters in the power system, consisting of $n_i = |S_{G_i}|$ TGs in Cluster i. We treat every generator as an "agent," and each agent is assigned a unique ID. Denote the generated power in Cluster i in a global vector as $P_i = \begin{bmatrix} P_i^1 & \cdots & P_i^j & \cdots & P_i^{n_i} \end{bmatrix}^T$. Taking the whole system into account, the global scheduled generated power for an m-cluster power system can be written as $P = \begin{bmatrix} P_1^T & P_2^T & \cdots & P_m^T \end{bmatrix}^T \in \mathbb{R}^n$, where $n = \sum\limits_{i=1}^{m} n_i$ is the total number of generators.

We assume that the demand supply constraint is a global constraint set, i.e.,

$$
X_g = \{P | \mathbf{1}^T P = P_d\}
\tag{7.4}
$$

where $\mathbf{1} \in \mathbb{R}^n$ is a $n \times 1$ vector with all the elements being 1, which is only known by the coordinator agent, an extra agent that is employed on top of all the local agents. The local constraint set is defined as the generator constraint, i.e.,

$$
X_i^j = \{P_i^j | P_i^{j\,\mathrm{min}} \leq P_i^j \leq P_i^{j\,\mathrm{max}}\}.
\tag{7.5}
$$

Denote the optimal solution of (7.3) as $P^\star \in \mathbb{R}^n$, which is unknown to each agent. Besides, each agent only has access to its own cost function f_i^j and constraint set X_i^j, which cannot be shared with other agents and in reality are the confidential information of the local TG. Our idea is that each agent estimates its own power output only based on its own information and proper coordination from the coordination agent. Denote the estimate of Agent j in Cluster i at iteration l as $\hat{P}_i^j(l)$, then we aim at proposing a scheme to achieve that $\lim\limits_{l \to \infty} \begin{bmatrix} \hat{P}_1^1(l) & \cdots & \hat{P}_i^j(l) & \cdots & \hat{P}_m^{n_m}(l) \end{bmatrix}^T = P^\star$.

7.3 Hierarchical Decentralized Economic Dispatch

7.3.1 Distributed algorithm in Chapter 5

In order to solve the ED problem formulated in (6.3), let us recall the distributed algorithm proposed in Chapter 5, where each generator estimates the global power output subject to both its own local constraint (generator constraint) and the global constraint (supply and demand constraint). The distributed algorithm is summarized as follows.

$$x^k(l+1) = P_{X_k}[z^k(l) - \zeta_l \nabla c_k(z^k(l))] \tag{7.6}$$

$$z^k(l) = \frac{\sum_{k=1}^{n} x^k(l)}{n} \tag{7.7}$$

where $P_{X_k}[.]$ is the projection operator, n is the number of total generators, $x^k(l) \in \mathbb{R}^n$ is the estimate of Agent k at iteration l, X_k is the intersection constraint set of both local constraint and global constraint. The average of all the estimates in $z^k(l)$ in (7.7) is obtained through a FACA algorithm in each iteration.

Note that this algorithm requires that each agent holds the global estimates, whose dimension is equal to the number of all the generator agents. When the system becomes larger and larger, both the computation and communication burden for each agent will increase. Besides, the projection constraint set in each iteration is the intersection of both generator constraint and global constraint, which makes the projection operation more complicated.

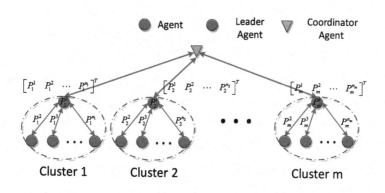

Figure 7.1: Hierarchical optimization structure

7.3.2 Hierarchical decentralized algorithm

In order to overcome the disadvantages mentioned above, in this chapter, a hierarchical decentralized algorithm involving three layers as illustrated in Fig. 7.1 is proposed based on (7.6) and (7.7). Firstly, we assign one agent as leader agent in each cluster. Without loss of generality, the leader agent is labeled as Agent 1 in each cluster.

Besides, except for these generator agents, an extra coordinator agent is employed to communicate with each leader agent.

All the generator agents (except for the leader agents) are in the first layer (blue circle), while the leader agents of respective clusters are in the second layer (brown circle). It is assumed that leader agent can communicate with all the other agents in the same cluster. On the top is the coordinator agent (green triangle), which mainly plays roles in two aspects, i.e., coordinating all the leader agents and handing the global constraints. It is also assumed that only the coordinator agent has access to the total load demand P_d.

In each iteration, the local generator agent only updates its own estimated power output and sends it to the leader agent. Thus the dimension of estimate and communication information for each agent decreases from n to 1, which is independent of the size of the whole system. Besides, only the local generator constraint projection is needed for each generator agent, which is much simpler. The global constraint is separately handled by the coordinator agent.

The proposed algorithm is illustrated in detail in Fig. 7.2, which is also summarized as follows.

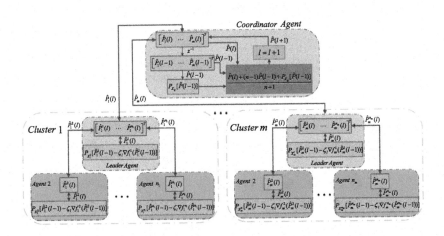

Figure 7.2: Hierarchical optimization structure for economic dispatch

Step 0 Initialization: As a starting point, at iteration $l = 0$, each local agent starts estimating its own scheduled power output $\hat{P}_i^j$ by choosing one arbitrary initial value $\hat{P}_i^j(0) \in \mathbb{R}$.

Step 1 Local optimization: At iteration step l, $l \geq 1$, each agent updates its local estimate $\hat{P}_i^j(l)$ by taking a gradient step to minimize its own cost function

f_i^j and then conducting a local constraint projection, as detailed below.

$$\hat{P}'^j_i(l) = \hat{P}^j_i(l-1) - \zeta_l \nabla f_i^j(\hat{P}^j_i(l-1)) \tag{7.8}$$

$$\hat{P}^j_i(l) = P_{X_i^j}[\hat{P}'^j_i(l)]$$

$$= \begin{cases} P_i^{j\max}, & \text{if } \hat{P}'^j_i(l) > P_i^{j\max} \\ P_i^{j\min}, & \text{if } \hat{P}'^j_i(l) < P_i^{j\min} \\ \hat{P}'^j_i(l), & \text{otherwise} \end{cases} \tag{7.9}$$

where ζ_l is a stepsize at iteration l, ∇f_i^j denotes the gradient of the cost function f_i^j.

Step 2   Cluster estimate: Each agent sends its local estimate $\hat{P}^j_i(l)$ to the leader agent. The leader agent forms a vector as a cluster estimate $\hat{P}_i(l)$, i.e., $\hat{P}_i(l) = \begin{bmatrix} \hat{P}_i^1(l) & \cdots & \hat{P}_i^j(l) & \cdots & \hat{P}_i^{n_i}(l) \end{bmatrix}^T \in \mathbb{R}^{n_i}$.

Step 3   Global constraint projection: The leader agent sends the cluster estimate $\hat{P}_i(l)$ to the coordinator agent. The coordinator agent obtains and saves the global estimate by forming all the cluster estimates as a vector, i.e., $\hat{P}(l) = \begin{bmatrix} \hat{P}_1(l)^T & \cdots & \hat{P}_i(l)^T & \cdots & \hat{P}_m(l)^T \end{bmatrix}^T \in \mathbb{R}^n$. Then the coordinator agent conducts the global constraint projection on the global estimates obtained in previous iteration step $l-1$ as follows.

$$P_{X_g}[\hat{P}(l-1)]$$

$$= \hat{P}(l-1) - \frac{\vec{N}^T \hat{P}(l-1) - P_d}{n} \vec{N} \tag{7.10}$$

where $\vec{N} = \begin{bmatrix} 1 & \cdots & 1 \end{bmatrix}^T \in \mathbb{R}^{n \times 1}$ is the normal vector to the surface of global constraint set X_g.

Step 4   Coordination operation: A coordination operation is conducted by the coordinator agent as follows.

$$\hat{P}(l+1)$$

$$= \frac{\hat{P}(l) + (n-1)\hat{P}(l-1) + P_{X_g}[\hat{P}(l-1)]}{n+1} \tag{7.11}$$

Step 5   Return updated estimate: The leader agent as well as the local agents in each cluster can receive the updated estimate from the coordinator agent as $\hat{P}^j_i(l+1) = [\hat{P}(l+1)]_{h_i^j}$, where $[.]_a$ denotes the ath element of the vector,

$h_i^j = \sum\limits_{s=1}^{i} n_{s-1} + j$ with $n_0 = 0$. Then go to Step 1.

Remark 7.1 *The proposed hierarchical decentralized algorithm in (7.8)-(7.11) is based on but significantly improved from the distributed algorithm in (7.6)-(7.7). The similarities and differences are summarized as follows. Firstly, similar to the local update in (7.6), the manipulations in (7.8) and (7.9) are also conducted in each local generator agent. But the dimension of the local update in (7.8) does not depend on the size of the whole system and it actually decreases from n to 1. Secondly, with the newly proposed algorithm, the local and global constraints are handled separately. Only local constraint projection is conducted in (7.9), and the global constraint is considered in the coordinator agent; while in (7.6), the projection constraint set contains both local constraint and global constraint. Clearly, such separation of operations greatly reduces the computation complexity for each local agent. Thirdly, the coordination manipulation in (7.11) is motivated by the average idea in (7.7).*

Remark 7.2 *One of the main purposes of clustering all the agents and assigning the leader agent in each cluster is to reduce the number of communication links between the local agent and coordinator agent. Besides, this structure may also bring some other potential benefits. For example, if there exist some cluster constraints in this cluster, the leader agent can handle these constraints by utilizing the dummy agent approach, an innovative idea conducted by the coordinator agent in (7.8), which will be explained in detail later.*

Remark 7.3 *It is worth pointing out that although the cost function formulated in (7.2) is quadratic, our proposed algorithm can also be applicable to any non-quadratic but convex cost function, such as that in Chapter 5.*

7.3.3 Convergence analysis

In this subsection, we analyze and prove the convergence of the proposed hierarchical decentralized algorithm. Firstly, we decompose the formulated problem in (7.3). Then by introducing the concept of dummy agent, which is one of the key ideas in our analysis, we will show that the convergence of the proposed hierarchical decentralized algorithm is equivalent to the distributed algorithm in Chapter 5 and [130]. Lastly the proof follows from the result in [16].

7.3.3.1 Problem decomposition

Recall that the global estimate for all the generators is $\hat{P}(l) \in \mathbb{R}^n$. The cost function and constraint set of Agent $k, k = h_i^j = \sum_{s=1}^{i} n_{s-1} + j$, $n_0 = 0$ with respect to $\hat{P}$ are denoted as follows, respectively:

$$c_k(P) = f_i^j([P]_{h_i^j}) \tag{7.12}$$

$$X_k = \{P | P_i^{j\min} \leq [P]_{h_i^j} \leq P_i^{j\min}\} \tag{7.13}$$

Then, (7.3) can be reformulated as

$$
\begin{cases}
\min_{P} \sum_{k=1}^{n} c_k(P) \\
s.t. \quad P \in \cap_{k=1}^{n} X_k \cap X_g
\end{cases}
\tag{7.14}
$$

It is noted that in (7.14), apart from that each agent has its own cost function and constraint set, there is one remaining global constraint set X_g. In order to merge this to get a more compact set, a novel new concept of dummy agent will be introduced in the following.

7.3.3.2 Dummy agent

The *dummy agent* proposed here is similar to the *virtual agent* in Chapter 6, which refers to an agent that has a constant cost function, i.e., no optimization variable, but subject to certain constraints. The objective cost function of the whole system will not change if we add such a dummy agent to the system.

Suppose we add one extra dummy agent into (7.14) and label it as Agent $n+1$. By setting the cost function and constraint set of the virtual agent as $c_{n+1} = 0, X_{n+1} = X_g$, (7.14) can be equivalently re-formulated as

$$
\begin{cases}
\min_{P} \sum_{k=1}^{n+1} c_k(P) \\
s.t. \quad P \in \cap_{k=1}^{n+1} X_k
\end{cases}
\tag{7.15}
$$

Theorem 7.1 Let $\{\hat{P}_i^j(l)\}, i = 1, \cdots, m, j = 1, \cdots, n_i$ be the estimates generated by the algorithm (7.8)-(7.11) and $X = \cap_{k=1}^{n+1} X_k$ be the intersection constraint set. Then the sequence $\{\hat{P}(l)\}$ converges to the optimal solution $P^\star$ with $P^\star \in X$, i.e.,

$$
\lim_{l \to \infty} \hat{P}(l) = P^\star
$$

if the stepsize ζ_l satisfies that $\zeta_l > 0$, $\sum_l \zeta_l = \infty$ and $\sum_l \zeta_l^2 < \infty$.

Proof : At iteration l, where $l \geq 1$,

$$\hat{P}(l) + (n-1)\hat{P}(l-1)$$

$$= \underbrace{\begin{bmatrix} \hat{P}_1^1(l) \\ \hat{P}_1^2(l-1) \\ \vdots \\ \hat{P}_m^{n_m}(l-1) \end{bmatrix}}_{\hat{x}_1^1(l)} + \underbrace{\begin{bmatrix} \hat{P}_1^1(l-1) \\ \hat{P}_1^2(l) \\ \vdots \\ \hat{P}_m^{n_m}(l-1) \end{bmatrix}}_{\hat{x}_1^2(l)} + \cdots$$

$$+ \underbrace{\begin{bmatrix} \hat{P}_1^1(l-1) \\ \hat{P}_1^2(l-1) \\ \vdots \\ \hat{P}_m^{n_m}(l) \end{bmatrix}}_{\hat{x}_m^{n_m}(l)} \qquad (7.16)$$

According to (7.8), (7.9), we have

$$\hat{x}_i^j(l) = \begin{bmatrix} \hat{P}_1^1(l-1) \\ \vdots \\ \hat{P}_i^j(l) \\ \vdots \\ \hat{P}_m^{n_m}(l-1) \end{bmatrix} = P_{X_i^j}[\hat{P}(l-1) - \zeta_l \nabla f_i^j(\hat{P}(l-1))]$$

$$i = 1, \cdots, m, j = 1, \cdots, n_i \qquad (7.17)$$

Substituting (7.17) into (7.16), we have

$$\hat{P}(l) + (n-1)\hat{P}(l-1)$$

$$= \sum_{i=1}^{m} \sum_{j=1}^{n_i} P_{X_i^j}[\hat{P}(l-1) - \zeta_l \nabla f_i^j(\hat{P}(l-1))]$$

$$= \sum_{k=1}^{n} P_{X_k}[\hat{P}(l-1) - \zeta_l \nabla c_k(\hat{P}(l-1))] \qquad (7.18)$$

Let us consider the dummy agent (Agent $n+1$). As $c_{n+1} = 0$, hence $\nabla c_{n+1} = 0$, then we have

$$P_{X_{n+1}}[\hat{P}(l-1) - \zeta_l \nabla c_{n+1}(\hat{P}(l-1))]$$

$$= P_{X_{n+1}}[\hat{P}(l-1)] = P_{X_g}[\hat{P}(l-1)] \qquad (7.19)$$

Combining (7.18) and (7.19), we have

$$\hat{P}(l) + (n-1)\hat{P}(l-1) + P_{X_g}[\hat{P}(l-1)]$$

$$= \sum_{k=1}^{n+1} P_{X_k}[\hat{P}(l-1) - \zeta_l \nabla c_k(\hat{P}(l-1))] \qquad (7.20)$$

Substituting (7.20) into (7.11), we have

$$\hat{P}(l+1) = \frac{\sum\limits_{k=1}^{n+1} P_{X_k}[\hat{P}(l-1) - \zeta_l \nabla c_k(\hat{P}(l-1))]}{n+1} \tag{7.21}$$

By further exploring (7.21), it gives the average of all the $n+1$ agents' estimates, which is equivalent to the distributed algorithm in (7.6)-(7.7) and in Proposition 5 in [130]. As the problem formulated in (7.15) is also equivalent to that in [130], then the remaining proof follows from that in [130]. ■

Remark 7.4 *From the above analysis, it is seen that the main purpose of introducing the concept of dummy agent is to handle the global constraint (demand supply equality constraint). In fact, this idea can be further employed to decompose some complex constraint set into the intersection of several simple constraint sets, where each simple constraint set can be represented by one dummy agent.*

From the above analysis, (7.11) corresponds to that only one dummy agent is introduced in the analysis. In fact, the number of dummy agents can be chosen as any nonzero integer α by modifying (7.11) as follows.

$$\hat{P}(l+1) = \frac{\hat{P}(l) + (n-1)\hat{P}(l-1) + \alpha P_{X_g}[\hat{P}(l-1)]}{n+\alpha} \tag{7.22}$$

Obviously, the more dummy agents there are, the faster $\hat{P}$ will converge to the constraint set X_g. On the other hand, if $\alpha = 0$, i.e., no dummy agent is included; then the global constraint set X_g will never be met. In other words, such a dummy agent is necessary to handle the global constraint. In the case studies of the next section, we will investigate the impact on algorithm convergence with different values of α.

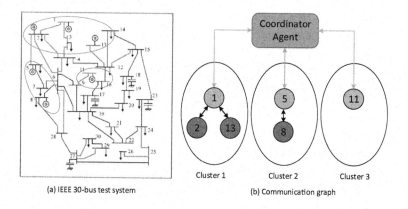

(a) IEEE 30-bus test system

(b) Communication graph

Figure 7.3: IEEE 30-bus test system

7.4 Case Studies

In order to test the effectiveness of the proposed hierarchical decentralized ED method, several case studies are presented and discussed in this section. The IEEE 30-bus system is used as a test system. Firstly, the implementation without and with generator constraints is demonstrated. The plug-and-play property of the proposed algorithm including both generator and load is illustrated. Then the impact on algorithm convergence with respect to different α is studied. Lastly, the IEEE 118-bus power system is employed as a large-scale power system case.

7.4.1 IEEE 30-bus system

In this subsection, the IEEE 30-bus system is chosen as a test system. The generator and load bus parameters are adopted from [90], which are the same as in Table 5.5 and Table 5.6, respectively. First, the 6 generators are divided into 3 clusters according to the distance of their physical connections, i.e., generators 1, 2, and 13 are in Cluster 1; generators 5 and 8 are in Cluster 2, and generator 11 is in Cluster 3, shown in Fig. 7.3. Without loss of generality, generators 1, 5, and 11 are assigned as the respective leader agents. Then the generators 1, 2, 13 are Agent 1, Agent 2, and Agent 3 in Cluster 1, while generators 5 and 8 are Agent 1 and Agent 2 in Cluster 2, generator 11 is Agent 1 in Cluster 3. The communication graph is shown in Fig. 7.3 (b). Initially it is easy to obtain from Table 5.6 that the total load demand

$$P_d = \sum_{i=1}^{30} P_i^d = 283.4MW.$$ Several case studies are presented and discussed here. In the first 2 case studies, the situations without and with generator constraints are investigated.

7.4.1.1 Case study 1: Without generator constraints

In this case study, we first illustrate our algorithm on the situation where the generator constraints are not imposed, where (7.7) in Step 2 becomes $\hat{P}_i^j(l) = \hat{P'}_i^j(l)$. Each generator agent chooses the initial value as $\hat{P}_i^j(0) = 0, \forall i, j$, respectively. To meet the stepsize condition, a diminishing stepsize is chosen as $\zeta_l = \frac{\zeta_0}{l+1000}, l \geq 1$, where ζ_0 is the initial stepsize and is set as $\zeta_0 = 1000$. The simulation results are shown in Fig. 7.4. From Fig. 7.4 (a), we see that the estimates of each agent $\hat{P}_i^j(l)$ converge to the optimal point $P^* = \begin{bmatrix} 189.13 & 47.71 & 8.40 & 19.36 & 10.19 & 8.40 \end{bmatrix}^T$, which indicates the optimal power allocation for each generator is that in Cluster 1 $P_{G_1} = 189.13MW, P_{G_2} = 47.71MW, P_{G_{13}} = 8.40MW$, in Cluster 2 $P_{G_5} = 19.36MW, P_{G_8} = 10.19MW$, in Cluster 3 $P_{G_{11}} = 8.40MW$. Note that $P_{G_{11}}, P_{G_{13}}$ are in conflict with their lower bound $P_{11}{}^{min} = 10MW, P_{13}{}^{min} = 12MW$, respectively. To clearly demonstrate the result, the errors of estimates of each agent are detailed in Fig. 7.4 (b) with a horizontal logarithmic scale.

7.4.1.2 Case study 2: With generator constraints

In this case study, we consider the generator constraints, where all the initial values are the same as those in case study 1 except for considering the generator constraints. The simulation results are shown in Fig. 7.5. The optimized power outputs are that in Cluster 1 $P_{G_1} = 185.11MW$, $P_{G_2} = 46.87MW$, $P_{G_{13}} = 12.03MW$, in Cluster 2 $P_{G_5} = 19.12MW$, $P_{G_8} = 10.03MW$, in area 3 $P_{G_{11}} = 10.03MW$. In this case, all the generators' power output are within their constraints. This simulation is carried out using Matlab® R2014a on a PC with an Intel Core i7-4770 CPU (3.40GHz) and 8 GB RAM. It takes approximately 10s to reach the iteration number $l = 5 \times 10^4$, where all the agents almost converge to the optimal solution.

These two cases show that our proposed algorithm can handle the multi-area ED problem both without and with generator constraints.

7.4.1.3 Case study 3: Varying load

In order to demonstrate the suitability of our algorithm for practical economic dispatch, a 2-hour dynamic load profile is chosen, as shown in Fig. 7.6(a). This total load demand is generated randomly every 5 minutes with ±50% variation around the initial demand $P_d = 283.4MW$. The estimated optimal power output of each generator is shown in Fig. 7.6(b). It can be seen that under the proposed hierarchical decentralized optimization architecture, each generator agent is able to obtain the optimal solution under a dynamic load profile.

7.4.1.4 Case study 4: Plug-and-play capability

This case study is to test the flexibility of the proposed algorithm. The plug-and-play of a certain generator is considered here. The simulation results are shown in Fig. 7.7. Suppose generator 2 (Agent 2 in Cluster 1) is disconnected from the power system at the time step $l = 1 \times 10^5$. The remaining generators obtain new optimal power output $P_{G_1} = 200.00MW$, $P_{G_{13}} = 15.26MW$, $P_{G_5} = 22.11MW$, $P_{G_8} = 30.77MW$, $P_{G_{11}} = 15.26MW$. Then at the time step $l = 2 \times 10^5$, generator 2 is connected to the system again and all the results converge to those of the previous ones.

7.4.1.5 Case study 5: Comparison with different α values

As pointed out in Eqn. (7.22), our proposed algorithm can be modified with multiple dummy agents. In order to investigate the impact of the number of dummy agents, in this case study, we study the convergence speed and the global constraint satisfaction under different values of α. The simulation results are shown in Fig. 7.8. From Fig. 7.8 (a), we can see that the smaller the α, the faster the convergence speed to the optimal point is. While from Fig. 7.8 (b), it is observed that larger α leads to quicker satisfaction of the equality constraint X_g (the demand and supply constraint). As the final solution to the problem should not only converge to the optimal but also satisfy the constraint set, hence α needs to be chosen properly by considering both the convergence speed and the satisfaction of the constraint set. In this case study, a proper

α can be chosen as 10 based on the results in Fig. 7.8, which is a trade-off choice between the convergence speed and the satisfaction of the constraint set.

7.4.1.6 Case study 6: Fast gradient

Note that, in Step 1 of our algorithm, only one iteration information is used. Motivated by the accelerated gradient method in [175], we hope to speed up the convergence by adding a momentum term $\rho(l-1) = \hat{P}(l-1) - \hat{P}(l-2)$ in (7.11), i.e.,

$$\hat{P}(l+1) = \frac{\hat{P}(l) + (n-1)\hat{P}(l-1) + P_{X_g}[\hat{P}(l-1)]}{n+1}$$
$$+ \eta\rho(l-1) \qquad (7.23)$$

where η is called the speed-up gain and usually chosen as $0 < \eta < 1$. The algorithm with (7.23) replacing (7.11) is tested by simulation studies by setting different values of η in this case study. The simulation results are shown in Fig. 7.9. A logarithmic scale is used to show the difference of convergence speed more clearly, especially during the beginning of the evolution. Compared to the original one (solid line), it is observed that with the modification in (7.23) (dashed line), $\hat{P}(l)$ converges to the optimal value much faster and thus we also name the modified algorithm as fast gradient method. It is also observed that larger η makes the convergence speed faster but leads to some oscillations. However, establishing convergence property for this algorithm is challenging and thus still remains an open problem. Therefore it deserves further research.

7.4.2 IEEE 118-bus test system

In order to test the effectiveness of the proposed method for a large network, the IEEE 118-bus system is chosen as a test system in this case study. The cost function parameters of 54 generators are adopted from [177]. The algorithm initial conditions are chosen as in previous cases studies but with a total load demand of $P_d = 10000MW$. The simulation results are shown in Fig. 7.10. It is observed that the generator can estimate its own optimal power output with our proposed algorithm.

Acknowledgments

©2017 IEEE. Required, with permission, from Fanghong Guo, Changyun Wen, Jianfeng Mao, Jiawei Chen and Yong-Duan Song, "Hierarchical Decentralized Optimization Architecture for Economic Dispatch: A New Approach for Large-scale Power System," IEEE Transactions on Industrial Informatics, published online, DOI: 10.1109/TII.2017.2749264.

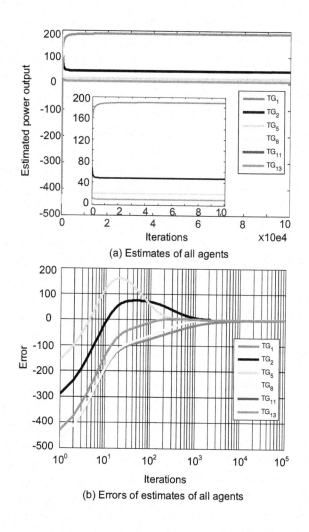

(a) Estimates of all agents

(b) Errors of estimates of all agents

Figure 7.4: Simulation results without generator constraints

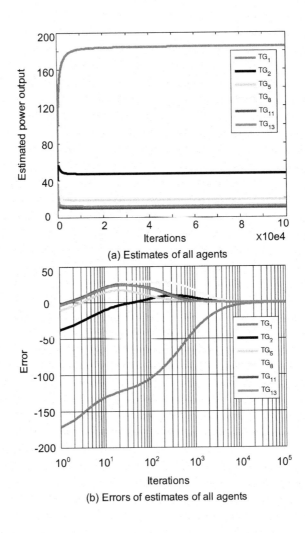

(a) Estimates of all agents

(b) Errors of estimates of all agents

Figure 7.5: Simulation results with generator constraints

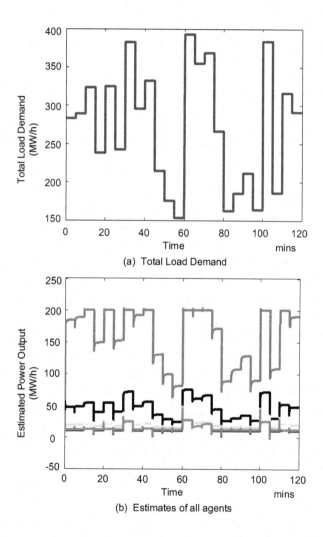

Figure 7.6: Simulation results of proposed algorithm with a 2-hour load profile

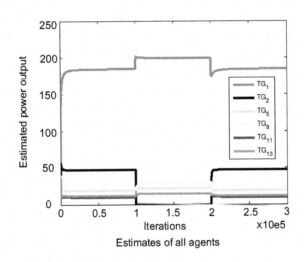

Figure 7.7: Simulation results with generator plug-and-play

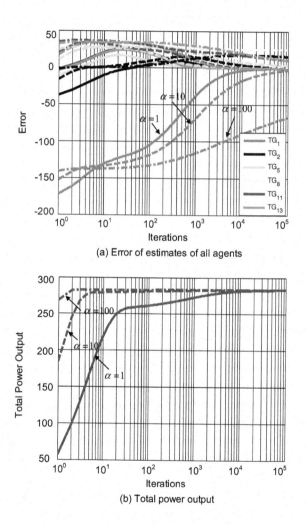

(a) Error of estimates of all agents

(b) Total power output

Figure 7.8: Simulation results with different α

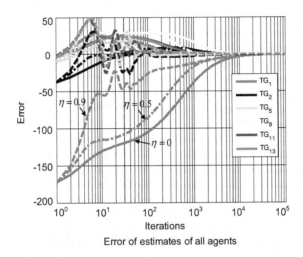

Error of estimates of all agents

Figure 7.9: Simulation results of fast gradient method

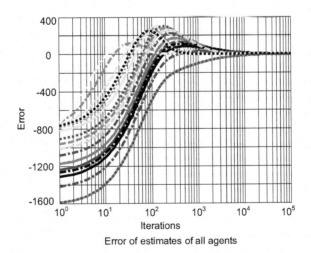

Error of estimates of all agents

Figure 7.10: Simulation results in IEEE 118-bus power system

Chapter 8

Distributed Optimal Energy Scheduling

In Chapters 5, 6, and 7, distributed optimization methods are developed to solve the economic dispatch problems in single-area and multi-area power systems, respectively. Motivated by the idea of distributed optimization, in this chapter, we consider another optimization problem in the tertiary control layer, i.e., the optimal energy scheduling problem, which is not a specific problem in MG control but more general in smart grid systems.

Pricing function plays an important role in the optimal energy scheduling problem in smart grid systems. In this chapter, we propose a novel real-time pricing strategy named proportional and derivative *(PD) pricing*. Different from conventional real-time pricing strategies, which only depend on the current total energy consumption, our proposed pricing strategy also takes the historical energy consumption into consideration, which aims to further fill the valley load and shave the peak load. An optimal energy scheduling problem is then formulated by minimizing the total social cost of the overall power system. Two different distributed optimization algorithms with different communication strategies are proposed to solve the problem. Several case studies implemented on a heating ventilation and air conditioning (HVAC) system are tested and discussed to show the effectiveness of both the proposed pricing function and distributed optimization algorithms.

8.1 Introduction

Optimal energy scheduling is a key problem in the electricity market for maintaining the balance between supply and demand [178]. Recently, with the advent of smart grid technologies, a number of information and communication infrastructures have been integrated into existing power systems, which enables real-time communication between the energy provider (supply side) and consumer (demand side) [179, 180]. Thus it offers an additional degree of freedom to do optimal energy management in the demand side. Demand side management (DSM) is envisioned as a key mechanism in the smart grid to effectively reduce the total energy costs and peak to average ratio (PAR) of the total energy demand [181]. The main target of DSM is to alternate the consumer's demand profile in time and/or shape, to make it match the supply [182].

In this chapter, inspired by the *PD* control law, one "derivative (D), i.e., the incremental value of total load demand, is added to the popular *P pricing* strategy, resulting in *PD pricing*. An optimal energy scheduling problem is formulated by minimizing the total social cost of a smart power system, motivated by the work in [105]. The power system considered here consists of a single energy provider, several loads and a regulatory authority. Each load and the energy provider are treated as individual local agents with different identities. Different from conventional centralized optimization algorithms, in this chapter, two different distributed optimization algorithms with two different communication strategies (synchronous and sequential communication) are proposed. The key idea behind distributed optimization is to decompose the central node into several sub-nodes (agents). In distributed optimization, each agent only accesses its local cost function and local constraint set, which protects the privacy of each local agent. By letting each local agent communicate with its neighboring agents, the global objective cost function can be minimized. Compared to centralized algorithms, a distributed one has the following major advantages: 1) less computation and communication cost; 2) plug-and-play property required by smart grid systems, which makes algorithm design more flexible; 3) robust to single-point-failures; and 4) easy and simple to design and implement as it only handles local information. Besides, it is worthy to point out that the formulated problem in this chapter is a coupled optimization problem, with coupling in both objective function and constraints. Some consensus-based optimization methods such as the distributed gradient method in [90] may fail to work.

8.2   Problem Formulation

8.2.1   System model

In this chapter, similar to [100], we consider a smart power system consisting of a single energy provider, several loads, and a regulatory authority, which is shown in Fig. 8.1. The loads in this smart power system are supposed to be equipped with an energy consumption controller (ECC) which is embedded in an advanced metering infrastructure (AMI). The role of ECC is to schedule the load's energy consumption, while the AMI is responsible for two-way communication among the loads and the regulatory authority. All the AMIs are connected to a local area network and each AMI can only communicate with its neighboring AMIs.

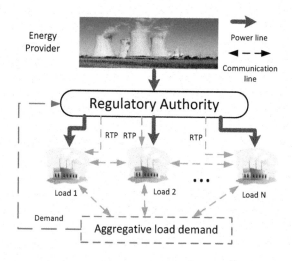

Figure 8.1: The smart power system

Suppose there are $N \triangleq |\mathcal{N}|$ loads, where $\mathcal{N}$ denotes the set of all loads. For each load $i \in \mathcal{N}$, let l_i^h denote the energy consumed by load i at time slot h, where $h \in \mathcal{H}$ with $\mathcal{H}$ being the set of all the time slots and a one-day operation cycle is divided into $H \triangleq |\mathcal{H}|$ slots. Similar to [15], the energy consumed by load i at time slot h is supposed to be bounded within an interval, i.e., $l_i^{h\min} \leq l_i^h \leq l_i^{h\max}$, where $l_i^{h\min}$ and $l_i^{h\max}$ denote the minimum and maximum energy consumption of load i at time slot h. Let L_h denote the generation capacity of the energy provider at time slot h. We assume that the energy provider has the minimum capacity to cover the minimum energy consumption requirement of all loads at each time slot, i.e., $L_h^{\min} \triangleq \sum_{i \in \mathcal{N}} l_i^{h\min} \leq L_h, \forall h \in \mathcal{H}$. We also impose the maximum capacity $L_h^{\max}$ to the energy provider, hence we have

$$L_h^{\min} \leq L_h \leq L_h^{\max}, \forall h \in \mathcal{H}. \tag{8.1}$$

8.2.2 Cost function

The utility function $U_i^h(l_i^h)$ is often employed to describe the consumers' comfort. Mathematically, $U_i^h(l_i^h)$ often appears in a quadratic or logarithmic function [183]. Then the total cost for loads (consumers) can be described as the combination of the load curtailment cost (negative utility function) and the electricity cost. The total cost for load i at time slot h can be formulated as [101], [105], [184]

$$C_i^h(\mathbf{l}^h) = -\mu_i^h U_i^h(l_i^h) + P(\mathbf{l}^h) l_i^h \tag{8.2}$$

where $0 < \mu_i^h \leq 1$ is a weight coefficient of the utility function, $\mathbf{l^h} = \begin{bmatrix} l_1^h & \cdots & l_N^h \end{bmatrix}^T$ is the aggregative energy consumption vector at time slot h, $P(\mathbf{l^h})$ is the real-time pricing function.

The cost function of the energy provider is usually approximated by a quadratic function [185]:

$$f_h(L_h) = \alpha_h L_h^2 + \beta_h L_h + \gamma_h \tag{8.3}$$

where $\alpha_h \in \mathbb{R}^+$, β_h and γ_h are the cost coefficients of the energy provider at time slot h.

8.2.3 Pricing function

Real-time pricing (RTP) is one of the key strategies in DSM. In most literatures, e.g., [101], [184], [186], the RTP function is defined as

$$P(\mathbf{l^h}) = p_1 \sum_{i \in \mathcal{N}} l_i^h + p_0 \tag{8.4}$$

where p_0 is a basic price for unit energy consumption, $p_1 \in \mathbb{R}^+$ is the coefficient of elastic pricing.

Obviously this RTP function is related to the total energy consumption of current time slot. The more energy consumed, the higher the price is. To some degree, this RTP strategy can help to remove the peak load by increasing the price and fill the valley load by decreasing the price. As pointed out in the introduction, this strategy is similar to a P controller, which only depends on the current energy consumption and is referred to as P pricing here. If the trend of energy consumption is also considered, it might improve the dynamic performance of the P pricing.

Suppose the curve of the total energy consumption of all loads in a one-day operation cycle is shown in Fig. 8.2. Ideally, the goal of demand-side management is to shape such a curve to the line of expected energy demand L^* (illustrated by the dashed line in Fig. 8.2) as close as possible. Such line divides the energy consumption curve into 2 periods. One intuitive idea is to use RTP as a tool to "encourage" users to increase their consumption during Period 1, while "punishing" them to reduce their consumption during Period 2. Also the more the incremental consumption changes, the more the price changes.

Hence motivated by the PD controller, an innovative "derivative" term is added to the P pricing strategy, i.e.,

$$P(\mathbf{l^h}) = p_1 \sum_{i \in \mathcal{N}} l_i^h + p_0 + p_2 \left(\sum_{i \in \mathcal{N}} l_i^h - L^{h-1} \right) \tag{8.5}$$

where $p_2 \in \mathbb{R}$ is the coefficient of incremental pricing, which can be designed by the history data, $L^{h-1} = \sum_{i \in \mathcal{N}} l_i^{h-1}$ is the total energy consumption at the time slot $h-1$.

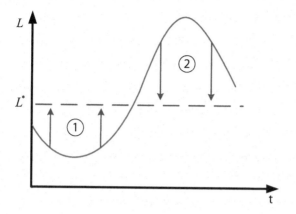

Figure 8.2: Illustration of *PD pricing* strategy

An example of the coefficient of incremental pricing p_2 can be designed as

$$p_2 = \begin{cases} p^+, & if \ L^{h-1} > L^* \\ -p^+, & if \ L^{h-1} < L^* \\ 0, & otherwise \end{cases} \qquad (8.6)$$

where p^+ is a positive number and L^* is the expected energy demand by regulatory authority.

Remark 8.1 *The RTP pricing strategy (8.5) is referred to as PD pricing. This strategy not only inherits the main pricing mechanism from P pricing, but also takes the incremental energy consumption into account. Specifically, if the total energy consumption of the last time slot is higher than the expected L^* and the total incremental energy consumption increases, the price will go much higher; otherwise the price will go much lower. Such a pricing strategy is expected to further fill the valley load and shave the peak load, and thus reduce the PAR of total load demand. Its performance and the impact of different parameter configuration will be illustrated and demonstrated in detail in the case study section.*

Note that the RTP strategy (8.5) is equivalent to (8.4) if we set $p_2 = 0$.

8.2.4 Optimization problem formulation

Considering the objective of achieving social optimum, i.e., to maximize the social welfare [101], or minimize the total social cost, the optimal energy scheduling problem is normally formulated as [100]

$$
\begin{cases}
\min\limits_{l_i^h, L_h} \sum\limits_{h\in\mathcal{H}} \sum\limits_{i\in\mathcal{N}} C_i(\mathbf{l}^h) + \sum\limits_{h\in\mathcal{H}} f_h(L_h) \\
s.t. \sum\limits_{i\in\mathcal{N}} l_i^h \le L_h, \forall h\in\mathcal{H} \\
l_i^{h\min} \le l_i^h \le l_i^{h\max}, \forall h\in\mathcal{H} \\
L_h^{\min} \le L_h \le L_h^{\max}, \forall h\in\mathcal{H}
\end{cases}
\tag{8.7}
$$

where $C_i(\mathbf{l}^h)$ and $f_h(L_h)$ are given in (8.2), (8.3), respectively.

Remark 8.2 *Though each load consumer is self-interested in nature, they can be driven to coordinate if they can benefit from the coordination. Incentive provoking mechanisms, such as benefit sharing mechanisms can be designed to motivate the consumer to participate in the coordination [184].*

As the incremental pricing gain p_2 has already been designed at time slot $h-1$ and L^{h-1} in (8.5) is also recorded in the ECC before the optimization at the time slot h, hence they will not affect the solution to (8.7) and thus can be treated as constants. In this scenario, (8.7) can be solved independently for each time slot $h \in \mathcal{H}$. In other words, for a fixed time slot $h \in \mathcal{H}$, we have the following optimization problem, which is equivalent to (8.7).

$$
\begin{cases}
\min\limits_{l_i^h, L_h} \sum\limits_{i\in\mathcal{N}} C_i(\mathbf{l}^h) + f_h(L_h) \\
s.t. \sum\limits_{i\in\mathcal{N}} l_i^h - L_h \le 0 \\
l_i^{h\min} \le l_i^h \le l_i^{h\max} \\
L_h^{\min} \le L_h \le L_h^{\max}
\end{cases}
\tag{8.8}
$$

Lemma 8.1 *The problem formulated in (8.8) is a convex optimization problem if the coefficients of proposed PD pricing strategy in (8.5)-(8.6) satisfy that $p_1 + p_2 \ge 0$.*

Proof : Rewriting the proposed *PD pricing* strategy in (8.5), we have

$$
P(\mathbf{l}^h) = (p_1 + p_2)\sum\limits_{i\in\mathcal{N}} l_i^h + (p_0 - p_2 L^{h-1})
\tag{8.9}
$$

Denote the Hessian matrix of $\sum\limits_{i\in\mathcal{N}} P(\mathbf{l}^h) l_i^h$ as H. Then $H = Diag\{2(p_1 + p_2), \cdots, 2(p_1 + p_2)\}$, which is a semi positive definite matrix if $p_1 + p_2 \ge 0$. The load curtailment cost appearing in the first part of (8.2) is a negative utility function, which is usually a quadratic function [100]. In addition, the cost function of the energy provider in (8.3) is also a quadratic function. Hence the total cost function in (8.8) is a convex function. Besides, it is easy to conclude that the constraint sets in (8.8) are also convex. Thus the problem formulated in (8.8) is a convex optimization problem. ■

The problem formulated in (8.8) can be easily solved by conventional centralized algorithms such as the interior point method [173]. However, as pointed out in [187], [188], such centralized algorithm has some drawbacks such as high computational burden and cost, and may also suffer from single-point failure. In this chapter, distributed optimization methods are proposed to solve (8.8). Firstly, we treat each load $i \in \mathcal{N}$ and energy provider as an "agent," and each agent is assigned a unique ID. Without the loss of generality, we assign first N agents as the loads, followed by the energy provider as the $(N+1)^{th}$ agent. Rewrite the optimization variables $\mathbf{l}^h$ and L_h in a compact form as $\mathbf{x} = \begin{bmatrix} \mathbf{l}^h & -L_h \end{bmatrix}^T$, which is an $(N+1) \times 1$ vector. Then the cost function c_i and the constraint set M_i of Agent i are denoted as follows, respectively,

$$c_i(\mathbf{x}) = \begin{cases} C_i(\mathbf{l}^h), i = 1, \cdots, N \\ f(-[\mathbf{x}]_{N+1}), i = N+1 \end{cases} \tag{8.10}$$

$$M_i = \begin{cases} \begin{array}{l} l_i^{\min} \leq [\mathbf{x}]_i \leq l_i^{\max} \\ \mathbf{1_{N+1}}^T \cdot \mathbf{x} \leq 0 \end{array}, i = 1, \cdots, N \\ \begin{array}{l} -L^{\max} \leq [\mathbf{x}]_i \leq -L^{\min} \\ \mathbf{1_{N+1}}^T \cdot \mathbf{x} \leq 0 \end{array}, i = N+1 \end{cases} \tag{8.11}$$

where $[\mathbf{x}]_i$ denotes the ith element of vector $\mathbf{x}$, $\mathbf{1_{N+1}}$ is a $(N+1) \times 1$ vector with all the elements being 1. Note that in order to simplify the notation, we have dropped the superscript h in the above parameters.

Remark 8.3 *It is worthy to point out that in order to facilitate the mathematical derivation of the projection operation onto the constraint set M_i, a minus sign is added in the front of L in the optimization vector $\mathbf{x}$ so that M_i appears in a summation form of the whole vector $\mathbf{x}$, as shown in (8.10).*

Now (8.8) can be reformulated as

$$\begin{cases} \min_{\mathbf{x}} \sum_{i=1}^{N+1} c_i(\mathbf{x}) \\ s.t. \quad \mathbf{x} \in \cap_{i=1}^{N+1} M_i \end{cases} \tag{8.12}$$

Denote the optimal solution of (8.12) as $\mathbf{x}^\star$. It is easy to conclude that $\mathbf{x}^\star$ exists and is unique according to the well-known *extreme value theorem*. But it is unknown to each agent. In addition, both cost function $c_i(\mathbf{x})$ and constraint M_i of Agent i, $i \in \mathcal{N}$ are only known by agent i itself. Our idea is that each agent estimates the optimal solution by using the available information of its neighboring agent $m, m \in \mathcal{N}_i$ and itself iteratively.

8.3 Distributed Optimal Energy Scheduling

In this section, two different distributed optimization algorithms are proposed for different situations. The first one is based on finite-time average consensus and projected gradient method with synchronous communication. However, for the situation that the agents are sparsely and remotely distributed, the communication latencies are quite different when using synchronous communication strategy. Then in order to overcome such a drawback, a sequential communication-based distributed optimization algorithm is also proposed. These two methods can both solve the problem formulated in (8.12) with different communication strategies.

In this chapter, the communication graph among the agents is modeled as an undirected connected graph.

8.3.1 *Distributed algorithm with synchronous communication*

In this subsection, motivated by the work in [130], a distributed optimization with synchronous communication is proposed to solve (8.12). Denote the estimate of the agent i at iteration k as $\mathbf{x}^i(k)$, $i = 1, \cdots, N+1$, which is also an $(N+1) \times 1$ vector. Then our proposed scheme is to ensure that $\lim_{k \to \infty} \mathbf{x}^i(k) = \mathbf{x}^*$, $i = 1, \cdots, N+1$. In other words, the estimates of all agents reach consensus of the optimal solution asymptotically. The flowchart of this method is shown in Fig. 8.3.

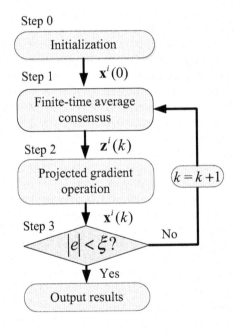

Figure 8.3: Flowchart of distributed optimization algorithm in each agent with synchronous communication

Unlike the distributed gradient method in [90], in Step 0, the initial value $\mathbf{x}^i(0)$, $i = 1, \cdots, N+1$ is allowed to be arbitrary.

In Step 1, the finite-time average consensus algorithm (FACA) introduced in Chapter 2 is applied. Its main idea is to let each agent communicate with its neighboring agents, and finally all the agents reach consensus to the average of their initial values in finite K steps, where K is determined by its predesigned communication graph. More specifically, K equals the number of distinct non-zero eigenvalues of the Laplacian matrix $\mathcal{L}$ [110]. The corresponding procedure is

$$
\left\{
\begin{aligned}
\mathbf{z}_1^i(k) &= w_{ii}(1)\mathbf{x}^i(k) + \sum_{j \in \mathcal{N}_i} w_{ij}(1)\mathbf{x}^i(k) \\
\mathbf{z}_2^i(k) &= w_{ii}(2)\mathbf{z}_1^i(k) + \sum_{j \in \mathcal{N}_i} w_{ij}(2)\mathbf{z}_1^j(k) \\
&\vdots \\
\mathbf{z}_K^i(k) &= w_{ii}(K)\mathbf{z}_{K-1}^i(k) + \sum_{j \in \mathcal{N}_i} w_{ij}(K)\mathbf{z}_{K-1}^j(k) \\
\mathbf{z}^i(k) &= \mathbf{z}_K^i(k)
\end{aligned}
\right.
\tag{8.13}
$$

where $w_{ii}(m)$, $w_{ij}(m)$, $m = 1, \cdots, K$ are the FACA updating gains. These gains can be obtained in a distributed fashion as

$$
w_{ij}(m) = \left\{
\begin{aligned}
&1 - \frac{n_i}{\lambda_m}, && j = i \\
&\frac{1}{\lambda_m}, && j \in \mathcal{N}_i \\
&0, && otherwise
\end{aligned}
\right. , m = 1, \cdots, K
\tag{8.14}
$$

where n_i is the number of the neighboring agents of agent i, $\lambda_m, m = 1, \cdots, K$ are distinct non-zero eigenvalues of the Laplacian matrix $\mathcal{L}$. More detailed information about how to obtain these gains can be found in [110].

In Step 2, each agent i updates its estimate by taking a gradient step to minimize its own cost function $c_i(\mathbf{x})$, and then projecting the result onto its constraint set M_i. This updating rule can be summarized as

$$
\mathbf{x}^i(k+1) = P_{M_i}\left[\mathbf{z}^i(k) - \zeta_k \nabla c_i(\mathbf{z}^i(k))\right]
\tag{8.15}
$$

where $P_{M_i}[.]$ is the projection operator described in Section 2.4.3, ζ_k is a stepsize at iteration k, ∇c_i denotes the gradient of the cost function c_i.

Theorem 8.1 *Let $\{\mathbf{x}^i(k)\}$, $i = 1, \cdots, N+1$ be the estimates generated by the algorithm (8.13)-(8.15) and $M = \cap_{i=1}^{N+1} M_i$. Then the sequence $\{\mathbf{x}^i(k)\}$, $i = 1, \cdots, N+1$ converges to the optimal solution $\mathbf{x}^\star$ with $\mathbf{x}^\star \in M$, i.e.,*

$$
\lim_{k \to \infty} \mathbf{x}^i(k) = \mathbf{x}^\star, i = 1, \cdots, N+1
$$

if the stepsize ζ_k satisfies that $\zeta_k > 0$, $\sum_k \zeta_k = \infty$ and $\sum_k \zeta_k^2 < \infty$.

Proof : Both the cost function $c_i(\mathbf{x})$ and constraint set M_i in (8.11) are convex, hence the convergence conclusion in Theorem 8.1 can be guaranteed by Theorem 5.1. ■

Remark 8.4 *As pointed out before, Algorithm 8.1 is mainly based on the synchronous communication, which is more suitable for the situation that the agents (loads and energy provider) are located closely. However, when the agents in the system are sparsely distributed, the communication latencies are quite different, such algorithm with the synchronous communication may not be proper. Hence in the next subsection, a sequential communication-based optimization method is proposed.*

8.3.2 *Distributed algorithm with sequential communication*

In this subsection, an alternative distributed optimization algorithm with sequential communication is proposed. The main idea is that each agent conducts its local optimization separately and in a sequential way. Motivated by the work in [133], we apply the Markov chain (MC) as the jump rule to determine the communication sequence.

Let $\mathbf{x}(k) \in \mathbb{R}^{N+1}$ be the estimate at iteration k. At each iteration, the estimate will be sent to only one agent, and this agent updates the estimate based on its own local information. Suppose currently Agent $w_k \in \mathcal{V}$, is conducting its local optimization, i.e.,

$$\mathbf{x}(k+1) = P_{M_{w_k}} \left[\mathbf{x}(k) - \zeta_k \nabla c_{w_k}(\mathbf{x}(k)) \right] \tag{8.16}$$

where w_k denotes which agent the estimate will be sent according to MC [133], which will be designed later.

Then it passes the estimate to one of its neighboring agents or just keeps it itself according to a transmission probability. The element of the transmission probability matrix of the MC $\mathcal{P}_{ij}, \forall i, j \in \mathcal{V}$ is set as [133]

$$\mathcal{P}_{ij} = \begin{cases} \min\left\{\frac{1}{n_i}, \frac{1}{n_j}\right\}, & j \in \mathcal{N}_i \\ 1 - \sum_{r \in \mathcal{N}_i} \min\left\{\frac{1}{n_i}, \frac{1}{n_r}\right\}, & j = i \\ 0, & otherwise \end{cases} \tag{8.17}$$

where n_i is the number of the neighboring agents of agent i.

Remark 8.5 *Note that $\mathcal{P}_{ij}$ is the transmission probability of Agent i to pass its estimate to Agent j. It is set by Agent i itself based on the number of the neighboring agents as well as its neighboring agents'. A distributed graph discovery algorithm based on "network flooding" proposed in Section 2.2.1 can be applied here for each agent to set $\mathcal{P}_{ij}$ in a distributed manner.*

Theorem 8.2 *Let $\{\mathbf{x}(k)\}$ be the estimates generated by the algorithm (7.16) according to the MC probability matrix designed in (8.17). Then the sequence $\{\mathbf{x}(k)\}$ almost surely converges to the optimal solution $\mathbf{x}^*$, if the stepsize ζ_k satisfies that $\zeta_k > 0, \sum_k \zeta_k = \infty$ and $\sum_k \zeta_k^2 < \infty$.*

Proof : It is easy to prove that the transmission probability matrix $\mathcal{P}_{ij}$ defined in (8.17) is a doubly stochastic matrix. According to Lemma 2 in [133], for a doubly stochastic transmission probability matrix, the expectations of the random number of visits to any state (agent) $i, i \in \mathcal{V}$ in the graph during a recurrence time are equal.

Also as the local constraint set M_i defined in (8.11) is a linear constraint set, it satisfies the *linear regularity* requirement in [189]. Hence the remaining proof procedure can follow from the proof of Proposition 5 in [189]. ■

To illustrate and compare these two optimization algorithms more clearly, a simple example consisting of 4 agents shown in Fig. 8.4 is used.

1) At each iteration, all the agents in Algorithm 8.1 conduct the local optimization and communicate with their neighboring agents synchronously; while in Algorithm 8.2, at each iteration, only one agent conducts the local optimization and passes its estimate to one of its neighboring agents including itself.

2) The communication among the agents in Algorithm 8.1 is fixed and deterministic once the communication graph is given; while in Algorithm 8.2, the communication between the agents is random, which is determined by the transmission probability matrix.

Taking Agent 2 in Fig. 8.4 (b) as an example, suppose at the current iteration, it receives the estimate from its neighboring agent (Agent 1 or itself) and conducts the local optimization based on (8.16). Afterwards it decides to which agent the updated estimate will be sent. According to (8.17), the elements of transmission matrix for Agent 2 are $\mathcal{P}_{21} = \mathcal{P}_{22} = \mathcal{P}_{23} = 1/3$, which means that the probabilities for Agent 2 to send its estimate to Agent 1, Agent 3 or keep it itself at next iteration are the same. To have a decision, a random variable in the range $[0, 1]$ is defined and generated randomly by computer. If it is in the range $[0, 1/3]$, then Agent 2 passes its estimate to Agent 1, if it is in the range $[1/3, 2/3]$, then passes to Agent 3; otherwise ,keeps it itself. So that the estimate will be sent to the 3 agents with equal probability.

3) The computational costs of both algorithms are almost the same (both consisting of one gradient descent step and one projection step), while the communication costs are different. At each iteration, Algorithm 8.1 has K steps FACA communication requirement internally, which has higher communication cost than that of Algorithm 8.2.

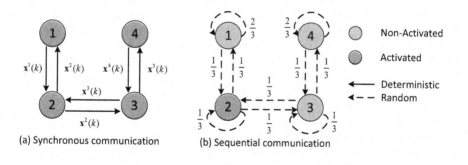

(a) Synchronous communication   (b) Sequential communication

Figure 8.4: Illustration example of two optimization algorithms

Remark 8.6 *Algorithm 8.1 is a synchronous algorithm, and is more suitable for the situation in which the agents are distributed closely. Algorithm 8.2 is an asynchronous algorithm, which can handle the situation where the agents are distributed sparsely. Compared to most existing distributed methods in [103]-[105], [185], [186] the proposed two algorithms are simple and easily implemented, as they are* *based on the projected gradient method. Also, different from the distributed gradient method in [90], the initial value of the proposed two algorithms is allowed to be arbitrary.*

8.4   Case Studies

In order to test the effectiveness of proposed *PD pricing* strategy as well as two distributed optimization methods, several case studies implemented on a HVAC system is considered in this section. It consists of 10 loads (consumers) and 1 energy provider, i.e., a total of 11 agents. The communication graph among the 11 agents is shown in Fig. 8.5. Firstly, the proposed two optimization algorithms are tested and compared at a particular time slot. Then the comparison between *P pricing* and *PD pricing* strategy is illustrated.

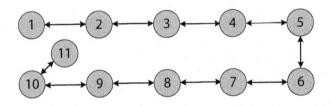

Figure 8.5: Communication graph

According to [101], [105], the utility function of HVAC load can be described as

$$U_i^h(l_i^h) = -\theta_i \gamma_i^2 (l_i^h - \hat{l}_i^h)^2 \tag{8.18}$$

where θ_i is the cost coefficient, γ_i is the parameter characteristic of the HVAC system, $\hat{l}_i^h$ is the expected amount of energy consumption of load i at time slot h. The parameters of HVAC loads are listed in Table 8.1.

8.4.1   Distributed optimization

In this section, the effectiveness of proposed optimization methods at a particular time slot h is tested. For an isolated time slot h, as there is no history data, hence $p_2 = 0$ in Eqn. (8.5), which is equivalent to (8.4). Note that in this case study, we assume that the maximum generating capacity $L^{\max}$ is equal to the sum of the maximum energy consumption of all users, i.e., $L^{\max} = \sum_{i=1}^{10} l_i^{\max}$.

Table 8.1: Parameters of HVAC system

Load i	Utility Function				
	$\theta_i \gamma_i^2$	μ_i	$\hat{l}_i$	$l_i^{\min}$	$l_i^{\max}$
1	0.90	0.4	100	70	100
2	0.91	0.4	110	75	110
3	0.92	0.4	120	80	120
4	0.93	0.4	130	85	130
5	0.94	0.4	140	90	140
6	0.95	0.4	150	95	150
7	0.96	0.4	160	100	160
8	0.97	0.4	170	105	170
9	0.98	0.4	180	110	180
10	0.99	0.4	190	115	190
Energy Provider	$\alpha = 0.0025,\ \beta = 0.03,\ L^{\min} = 925,\ L^{\max} = 1450$				
Pricing	$p_0 = 0.1,\ p_1 = 0.01$				

8.4.1.1 Distributed optimization with synchronous communication

Based on the communication graph given in Fig. 8.5, it is easy to calculate that it has 10 distinct nonzero eigenvalues, which means that it needs $K = 10$ steps for each agent to reach average consensus when applying FACA. As the initial value $\mathbf{x}^i(0)$ can be chosen arbitrarily, here we choose $\mathbf{x}^i(0) = \begin{bmatrix} 0 & \cdots & 0 \end{bmatrix}^T, i = 1, \cdots, 11$. To meet the stepsize condition, a diminishing stepsize is chosen as $\zeta_k = \frac{\zeta_0}{k}, k \geq 1$, where ζ_0 is the initial stepsize and is set as $\zeta_0 = 100$ in this case study.

The simulation results are shown in Fig. 8.6. From Fig. 8.6 (a), it is clear that the estimated energy consumption outputs of all agents $x^i(k)$, $i = 1, \cdots, 11$ almost reach consensus and get the optimal result $l_1 = 70kWh$, $l_2 = 75kWh$, $l_3 = 82.3kWh$, $l_4 = 92.8kWh$, $l_5 = 103.1kWh$, $l_6 = 113.5kWh$, $l_7 = 123.9kWh$, $l_8 = 134.3kWh$, $l_9 = 144.7kWh$, $l_{10} = 155kWh$, $L = 1094.6kWh$ after $k = 10 \times 10^3$ iterations. To clearly demonstrate details, the estimates of Agent 1 are shown in Fig. 8.6 (b). As expected, both individual energy consumption of each user $l_i, i = 1, \cdots, 10$ and the generating capacity L are bounded within the minimum and maximum energy requirements. Also note that the total energy consumption of all the users is equal to the generating capacity L, i.e., $\sum_{i=1}^{10} l_i = L$.

8.4.1.2 Distributed optimization with sequential communication

According to Eqn. (8.16) and the communication graph in Fig. 8.5, the MC transmission probability matrix $\mathcal{P}$ for sequential communication can be easily obtained as

$$\mathcal{P}_{ii} = \frac{2}{3}, \ \mathcal{P}_{ij} = \frac{1}{3}, \ i = 1, 11, \ j \in \mathcal{N}_i$$

$$\mathcal{P}_{ii} = \frac{1}{3}, \ \mathcal{P}_{ij} = \frac{1}{3}, \ i = 2, \cdots, 10, \ j \in \mathcal{N}_i$$

The simulation results are shown in Fig. 8.7 and 8.8. Initially, Agent 1 is chosen as the start agent with the initial value $\mathbf{x}(0) = \begin{bmatrix} 0 & \cdots & 0 \end{bmatrix}^T$. After conducting its local optimization, Agent 1 passes the estimate to Agent 2 or keeps it itself according to the designed MC transmission probability matrix. In this simulation, a random variable in the range [0, 1] is defined and produced randomly by computer. If the random variable is greater than 2/3, then Agent 1 sends the estimate to Agent 2; otherwise it keeps it itself.

The distribution of the agent which conducts the optimization is shown in Fig. 8.7. From Fig. 8.7 we can see that all the agents are almost visited equally often. The iteration process of estimating energy consumption $\mathbf{x}(k)$ is shown in Fig. 8.8, which converges to the same optimal values after $k = 10 \times 10^4$ iterations, the same as those in the previous optimization method with synchronous communication. However, it is also noted that the estimate trajectories of this algorithm are not as "smooth" as those in Algorithm 8.1 until the iteration $k = 5 \times 10^4$ because of its random communication strategy.

Motivated by the deterministic communication strategy in [132], we change the communication strategy in Algorithm 8.2 by applying the round-robin approach, as shown in Fig. 8.9 (a). The simulation results are shown in Fig. 8.9 (b). Compared to the results in Fig. 8.8, it takes fewer iterations to converge to the optimal solutions. Besides, the estimate trajectories are much more "smooth" than those in Fig. 8.8. However, one drawback of this communication strategy is that the communication graph needs to be re-designed to the one shown in Fig. 8.9 (a).

The above two case studies have shown the effectiveness of both our proposed distributed optimization algorithms.

8.4.2 Comparison between P pricing and PD pricing

The comparison between *P pricing* and *PD pricing* is now conducted. The entire time cycle is divided into 24 time slots representing 24 hours in a one-day cycle. According to the energy consumption trends in [101], [190], the weight coefficients of the utility function $\mu_i = \mu, \forall i = 1, \cdots, 10$ in different time slots are listed in Table 8.2. The parameters of the utility cost function for each load and the parameters of the cost function for the energy provider are supposed to be unchanged in all of the time slots and are the same as those listed in Table 8.2. For our proposed *PD pricing* in (8.5) and (8.6), the expected energy demand is set as $L^* = 1100kWh$ and the value of p^+ is chosen to be 0.005.

The results are shown in Fig. 8.10. The peak time is defined from $h = 18$ to $h = 22$. The total energy consumption during the peak time is reduced from $6.316 \times 10^3 kWh$ to $6.158 \times 10^3 kWh$ and the value of PAR is reduced from 23.4%

Table 8.2: Utility function coefficient in a one-day cycle

h	μ	h	μ	h	μ	h	μ
1	0.35	7	0.27	13	0.54	19	0.95
2	0.29	8	0.31	14	0.6	20	1
3	0.26	9	0.34	15	0.55	21	0.95
4	0.23	10	0.36	16	0.61	22	0.78
5	0.21	11	0.4	17	0.75	23	0.62
6	0.23	12	0.46	18	0.85	24	0.47

to 22.8% with the total energy consumption for the one-day cycle varying from $2.6989 \times 10^4 kWh$ to $2.7015 \times 10^4 kWh$ by using the proposed *PD pricing* strategy. Also from Fig. 8.10, it is observed that the total energy consumption of *PD pricing* during all of the time slots is much closer to the expected value (the black line) than those of *P pricing*. This shows that our proposed *PD pricing* strategy has the ability to further fill the valley load and shave the peak load.

It is also noted that there exists a large deviation from the time slot $h = 10$ to $h = 13$. The reason is that energy consumption during this period is close to the expected value, resulting in the derivative gain p_2 changing its sign frequently according to (8.6), which may have a large impact on the users' consumption. In order to evaluate such impact, the performances of different p^+ values are tested. The results are shown in Fig. 8.11. It is observed that with small derivative gain (Fig. 8.11 (a)), the performance of proposed *PD pricing* strategy is very similar to that of *P pricing* but with small improvement in valley load filling and peak load shaving. Meanwhile, if the derivative gain is chosen much larger, as shown in Fig. 8.11 (b), our proposed *PD pricing* strategy will have obvious impact on the costumers' power consumption, which makes the total energy consumption during the peak time be reduced from $6.316 \times 10^3 kWh$ to $6.0674 \times 10^3 kWh$. However, a larger derivative gain also causes a larger energy consumption deviation from the time slot $h = 9$ to $h = 12$. Hence, similar to parameter choice of conventional PD controller, in practice, a proper derivative gain should be carefully determined by the regulatory authority.

Acknowledgments

©2017 IEEE. Required, with permission, from Fanghong Guo, Changyun Wen, Zhengguo Li, "Distributed Optimal Energy Scheduling Based on a Novel PD Pricing Feedback Strategy in Smart Grid," 2015 IEEE 10th Conference on Industrial Electronics and Applications (ICIEA), pp. 208 -213, Auckland, New Zealand, 2015.

©2017 IET. Required, with permission, from Fanghong Guo, Changyun Wen, Zhengguo Li, "Distributed optimal energy scheduling based on a novel PD pricing strategy in smart grid," IET Generation, Transmission & Distribution, vol. 11, no. 8, pp. 2075-2084.

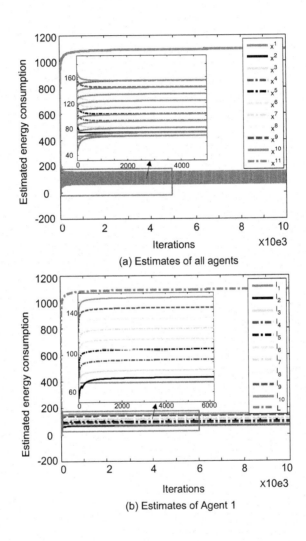

(a) Estimates of all agents

(b) Estimates of Agent 1

Figure 8.6: Simulation results of distributed optimization with synchronous communication

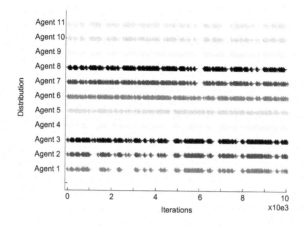

Figure 8.7: Distribution of agents conducting optimization during sequential communication process

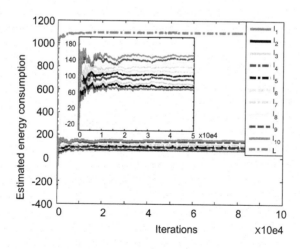

Figure 8.8: Simulation results of distributed optimization with random sequential communication

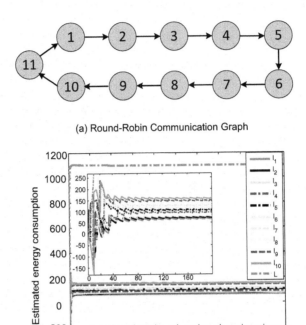

(a) Round-Robin Communication Graph

(b) Deterministic

Figure 8.9: Simulation results of distributed optimization with deterministic sequential communication

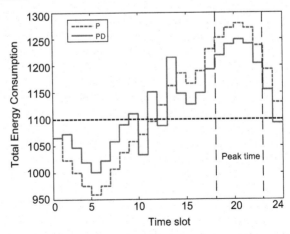

Figure 8.10: Comparison between *P* and *PD pricing* strategy

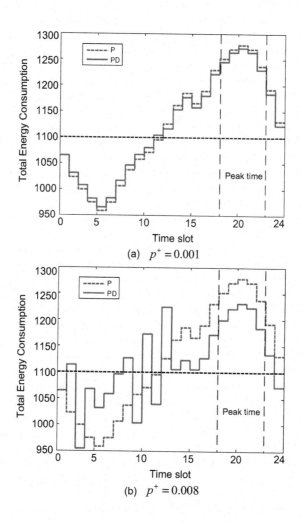

(a) $p^+ = 0.001$

(b) $p^+ = 0.008$

Figure 8.11: The performance of different derivative gains

Chapter 9

Conclusion and Future Works

9.1 Conclusion

In this book, we have investigated distributed control and optimization technologies for MG systems. The obtained results are summarized as follows:

1. The problem of voltage and frequency restoration in islanded microgrid systems is considered in Chapter 3. A distributed secondary control scheme is proposed to solve this problem. Compared to the existing centralized control approach, our method is fully distributed. In order to design and analyze the voltage and frequency secondary restoration control separately, we first apply the distributed finite-time approach to design the voltage controller, which enables the voltage regulation within finite time. Then the frequency restoration is addressed while keeping the real power sharing accuracy. A sufficient local stability condition for the proposed controller is given. Simulation results show that the proposed controller can restore the voltage and frequency of the whole system to their reference values while keeping a good real power sharing accuracy, regardless of whether additional load is connected to or disconnected from the system.

2. In order to overcome some intrinsic disadvantages brought by the centralized control, we propose a distributed secondary control scheme for voltage unbalance compensation in Chapter 4. The main key here is to design a distributed control scheme which seeks global information by allowing each local controller to communicate with its neighboring controllers. The proposed control scheme has a distributed two-layer secondary compensation architecture. This

architecture involves a finite time average-consensus algorithm and a newly developed graph discovery algorithm. The proposed control scheme is not only able to compensate well for the unbalanced voltage in SLB but also share the compensation effort dynamically in the distributed fashion. Also compared to the centralized control, this scheme has higher communication fault tolerance ability and has the plug-and-play property. A simulation test system is built in MATLAB® and several case studies have validated the effectiveness of our proposed strategy. Future work may include VUC in more general cases, such as 1) SLB is located in an arbitrary bus in a mesh topology; 2) transmission line with unbalanced impedance.

3. In Chapter 5, a fully distributed ED algorithm based on finite-time average consensus and projected gradient is proposed for smart grid systems with random wind power. By allowing each agent to communicate with its neighbor agents, the total cost of the whole system can be minimized by the proposed distributed ED algorithm while satisfying both equality and inequality constraints. Compared to the existing methods, no private confidential gradient or incremental cost information exchange is needed and the objective function is not required to be quadratic. What's more, the initial estimate values of our proposed method can be chosen arbitrarily by each agent individually. The effectiveness of the proposed scheme has been validated by several case studies including without generator constraints, with generator constraints, plug-and-play of generators and loads, and a large IEEE 30-bus test system. The results show good performance of the proposed method. As an alternative approach, how to develop a distributed ED strategy based on a stochastic programming method is an interesting topic worthy of consideration as a future work.

4. In Chapter 6, a distributed optimization algorithm is proposed for the multi-agent system which consists of multi groups of agents. The goal for the agents is to cooperatively minimize the sum of all the local cost functions, each of which is only known by the local agent itself. Also, each agent is constrained by a global and a local constraint set. Our proposed algorithm allows the agents in the same group to estimate the optimal solution individually and communicate with their neighboring agents in parallel. Once their estimates reach consensus, the leader agent in the group sends the estimate to the leader agent in a neighboring group in a sequential way. Two kinds of communication strategies, i.e., deterministic and random, are designed for the leader agents. The convergence of the proposed algorithm is theoretically proved with the virtue of the virtual agent. We also apply this algorithm to solve the economic dispatch problem in multi-area power systems. The effectiveness of the proposed algorithm has been validated by several case studies on IEEE 30-bus power systems. In order to accelerate the convergence speed, future work may consider modifying our algorithm with a fast gradient method.

5. In Chapter 7, we present a hierarchical decentralized optimization architecture for economic dispatch in smart grid. Different from conventional centralized ED methods, we decompose such problem into several sub-problems, which are solved by each local generator by using only its own information. In order to reduce the communication links, we divide the whole generator agents into clusters and assign a leader agent to each cluster. The leader agent gathers the whole cluster estimates and sends them to an extra coordinator agent. The main role of the coordinator agent is to handle the global demand and supply constraint as well as to coordinate among all the leader agents. It is theoretically shown that the proposed algorithm can ensure each local generator agent obtains the optimal power output when the chosen stepsizes are diminishing under certain conditions. The effectiveness of the proposed algorithm has been validated by several case studies implemented on IEEE 30-bus and IEEE 118-bus power systems. The results show good performance of the proposed method.

6. In Chapter 8, a distributed optimal energy scheduling strategy is proposed for a smart power system, which consists of a single energy provider, several loads and a regulatory authority. A novel real-time pricing strategy named *PD pricing* is proposed by adding an incremental term in the conventional *P pricing*. Two different distributed optimization methods are proposed to minimize the total social cost and they can be applied in different situations. The effectiveness of the proposed methods as well as the *PD pricing* strategy have been validated by case studies implemented on a HVAC system. It is worth noting that the proposed *PD pricing* strategy can be implemented in real applications with the development of big data technology [191]. Based on such technology, more information will be available for temporal and/or spatial integration of an ecosystem in practice. Subsequently, the overall performance of the ecosystem could be improved.

9.2   Recommendations for Future Research

Besides the work achieved in the literature and this book, there are still several interesting research topics that are worthy to be explored in the area of distributed control of energy systems (mainly MG systems), as outlined below.

9.2.1   Distributed adaptive control of the MG

In Chapter 3, we design the controllers under the assumption that the system parameters can be directly accessed and remain unchanged. However, in reality, there are some time varying parameters in the MG. For example, the installation of the power factor correction (PFC) capacitors can directly affect the transmission line's parameters. Also, the variation of the system topology (when the MG is reconfigured) will change the impedance parameters as seen by the DG units. Besides, there exist some unknown loads that are connected or disconnected to/from the MG load bus randomly, which may be treated as unknown external disturbances.

With the aforementioned uncertain system parameters and unknown external disturbances, the conventional PI or PR controllers used in the primary control layer may not work well. Some advanced control methods are needed, among which adaptive control is one of the promising methods. In addition, the interactions among the DG units are also not considered in the previous primary control because the overall dynamic cannot be obtained by each individual controller. Motivated by the above considerations, in the future research we may design distributed adaptive controllers in the primary control layer. Similar to [50], the MG system can be modeled by a dynamic equivalent circuit with uncertain parameters and unknown disturbances shown in Fig. 9.1. The main idea of developing such a control strategy is as follows: The uncertain parameters and the unknown external disturbances in the system can be estimated by adaptive laws, while the effects of the interactions between the DG units can be mitigated by using the neighboring DGs' information which can be obtained through the communication.

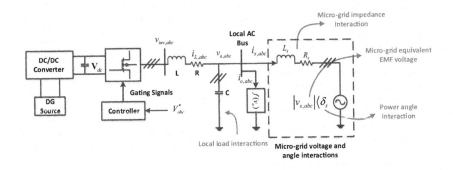

Figure 9.1: The dynamic equivalent circuit model of the MG system [50]

9.2.2 Stability analysis of the distributed secondary control with communication time delay

In Chapter 3 and 4, we have proposed the distributed secondary control schemes for the frequency restoration and voltage unbalance compensation. However, zero communication time delay has been assumed there. In reality, there always exist certain time delays in the communication. How to evaluate such delays and analyze the stability of the system under such situations becomes another research topic in our future work. Existing stability analysis with time delay can be divided into two classes: one is the frequency domain method; another is the time domain method.

The first method mainly involves the evaluation of the roots of the characteristic equation of small-signal model. This method is in principle exact but hard to use for the difficulty in determining the roots of the characteristic equation [192]. The time domain method is mainly based on Lyapunov-Krasovskii's stability theorem and Razumikhin's theorem [193]. This method can deal with uncertainties and time-varying delays. However, it cannot be used to find the delay stability margin. Also it strongly depends on the ability of defining a Lyapunov function.

One suggested way to handle this problem is based on the algorithm proposed in [194], which describes the impact of time-delays on small-signal angle stability of power systems. The proposed algorithm allows estimating an approximate solution of the characteristic equation of the delay differential algebraic equations based on the Chebyshev's differentiation matrix. It is able to precisely estimate the rightmost eigenvalues at an acceptable computational cost. We can employ this method to analyze the stability of the distributed secondary control with the communication time delays and also try to find the delay stability margin.

9.2.3 *Distributed event-triggered optimization of the MG*

As reviewed in Chapter 1, distributed control strategy has been widely used in the optimal power flow and economic dispatch problem of the MG, which is in the third layer according to the hierarchical definition of the MG control. Most of the existing studies are based on the continuous periodical information communication, which requires that the communication network have a large throughput capacity. However, in reality, the networks used in the MG are usually powered by batteries or solar arrays, which leads to a very limited throughput capacity. If we are able to limit the amount of exchange information between the controllers, the communication burden and cost of the network could be greatly decreased. One of the promising ways to do so is to adopt an event-triggered approach for the information transmission. Different from the conventional time-triggered system, the event-triggered system is the system in which all activities are activated by the occurrence of significant events. Usually these activities happen in a sporadic non-periodic manner.

In future work, motivated by the recent works in [91], [195], [196], the event-triggered strategy for communication in solving the ED problems in Chapters 5 and 6 is recommended. In fact, as pointed out before, the control problem at this layer has a quite slow dynamic response and does not require frequent information exchange. The controller only needs to transmit information to its neighboring controllers when a local error signal exceeds a specified threshold.

References

[1] A. M. Annaswamy, "IEEE vision for smart grid controls: 2030 and beyond," *IEEE Smart Grid Research*, 2013

[2] California Public Utilities Commission, "California Renewables Portfolio Standard (RPS)," [Online], Available

[3] European Commission, "Energy roadmap 2050," [Online], Available

[4] U.S. - Canada Power System Outage Task Force, Tech. Rep., "Final report on the August 14, 2003 blackout in the United States and Canada: Causes and recommendations," [Online], Available

[5] M. Yazdanian, A. Mehrizi-Sani, "Distributed control techniques in microgrids," *IEEE Transactions on Smart Grid*, vol. 5, no. 6, pp. 2901 - 2909, 2014

[6] P. D. Christofides, R. Scattolini, David Munoz, J. Liu, "Distributed model predictive control: A tutorial review and future research directions," *Computers and Chemical Engineering*, vol. 51, pp. 21 - 41, 2013

[7] F. Wang, "Agent-based control for networked traffic management systems," *IEEE Intelligent Systems*, vol. 20, no. 5, pp. 92 - 96, 2005

[8] Jan van de Vyver, G. Deconinck, R. Belmans, "The need for a distributed algorithm for control of the electrical power infrastructure," *Proceedings of International Symposium on Computational Intelligence for Measurement Systems and Applications*, pp. 211 - 215, Lugano, Switzerland, 2003

[9] U.S. Department of Energy., "Smart Grid / Department of Energy."

[10] X. Yu, C. Cecati, T. Dillon and M. Simoes, "The new frontier of smart grids: An industrial electronics perspective," *IEEE Industrial Electronics Magzine*, vol. 5, no. 3, pp. 49 - 63, September 2011

[11] X. Yu, Y. Xue, "Smart Grids: A cyber-physical systems perspective," *Proceedings of the IEEE*, vol. 104, no. 5, pp. 1058 - 1070, 2016

[12] R. H. Lasseter, "Microgrids," *in Proceedings of Power Engineering Scociety Winter Meeting*, vol. 1, pp.305 - 308, Jan. 2002

[13] R. H. Lasseter, "Smart Distribution: Coupled Microgrids," *Proceedings of the IEEE*, vol. 99, no. 6, pp. 1074 - 1082, 2011

[14] D. E. Olivares, A. Mehrizi-Sani, A. H. Etemadi and etc., "Trends in microgrid control," *IEEE Transactions on Smart Grid*, vol. 5, no. 4, pp. 1905 - 1919, 2014

[15] C. K. Sao ans W. Lehn, "Control and Power Management of Convert Fed Microgrids, " *IEEE Transactions on Power Systems*, vol. 23, no. 3, pp. 1088 - 1098, August 2008

[16] X. Liu, P. Wang and Poh Chiang Loh, "A hybrid AC/DC microgrid and its coordination control," *IEEE Transactions on Smart Grid*, vol. 2, no. 2, pp. 278 - 286, 2011

[17] M. E. Baran, N. R. Mahajan, "DC distribution for industrial systems: Opportunities and challenges," *IEEE Transactions on Industrial Appllication*, vol. 39, no. 6, pp. 1596 - 1601, November, 2003

[18] D. J. Hammerstrom, "AC versus DC distribution systems- Did we get it right?" *in IEEE Power Engineering Society General Meeting*, pp. 1 - 5, June, 2007

[19] About REIDS. [Online], Available

[20] Pulau Ubin Micro-grid Test-Bed. [Online], Available

[21] N. Hatziargyriou, H. Asano, R. Iravani, and C. Marnay, "Microgrids: An overview of ongoing research, development, and demonstration projects," *IEEE Power and Energy Magazine*, vol. 5, no. 4, pp. 78 - 94, 2007

[22] N. Hatziargyriou, "Microgrids: Large scale integration of micro-generation to low voltage grids," *1st International conference on Integration of RES and DER*, Brussels, 1 - 3, December 2004

[23] R. Lasseter, J. Eto, B. Schenkman, J. Stevens, H. Vollkommer, D. Klapp, E. Linton, H. Hurtado, and J. Roy, "Certs microgrid laboratory test bed, and smart loads," *IEEE Transactions on Power Delivery*, vol. 26, no. 1, pp. 325 - 332, January, 2011

[24] P. Tenti, A. Costabeber, P. mattavelli, D. Trombetti, "Distribution loss minimization by token ring control of power electronic interfaces in residential micro-grids," *IEEE Transactions on Industrial Electronics*, vol. 59, no. 10, pp. 3817 - 3826, 2012

[25] S. Chakraborty, M. D. Weiss, M. G. Simoes, "Distributed intelligent energy management system for a single-phase high-frequency AC microgrid," *IEEE Transactions on Industrial Electronics*, vol. 54, no. 1, pp. 97 - 109, 2007

[26] J. A. P. Lopes, C. L. Moreria and A. G. madureira, "Defining control strategies for microgrids islanded operation," *IEEE Transactions on Power System*, vol. 21, no. 2, pp. 916 - 924, May 2006

[27] J. C. Vasquez, J. M. Guerrero, J. Miret, M. Castilla, L. G de Vicua, "Hierarchical control of intelligent microgrids," *IEEE Industrial Electronics Magzine*, pp. 23 - 29, December, 2010

[28] J. M. Guerrero, M. Chandorkar, T. Lee, and P. C. Loh, "Advanced control architectures for intelligent microgrids-part I: Decentralized and hierarchical control," *IEEE Transactions on Industrial Electronics*, vol. 60, no. 4, pp. 1254 - 1262, 2013

[29] J. M. Guerrero, J. C. Vasquez, J. Matas, and L. G de Vicuna, "Hierarchical control of droop-controlled AC and DC microgrids - A general approach toward standardization, " *IEEE Transactions on Industrial Electronics*, vol. 58, no. 1, pp. 158 - 172, January, 2011

[30] A. Bidram and A. Davoudi, "Hierarchical structure of microgrids control system," *IEEE Transactions on Smart Grid*, vol. 3, no. 12, pp. 1963 - 1976, December, 2012

[31] A. S. Dobakhshari, S. Azizi and A. M. Ranjbar, "Control of microgrids: Aspects and prospects," *in Proceedings of International Conference on Network, Sensors and Control*, pp. 38 - 43, April, 2011

[32] F. Guo, C. Wen, "Distributed control subjected to constraints on control inputs: A case study on secondary control of droop-controlled inverter-based microgrids," *2014 IEEE 9th Conference on Industrial Electronics and Application*, Hangzhou, China, pp. 1119 - 1124, 2014

[33] F. Guo, C. Wen, J. Mao, and Y.-D. Song, "Distributed secondary voltage and frequency restoration control of droop-controlled inverter-based microgrids," *IEEE Transactions on Industrial Electronics*, vol. 62, no. 7, pp.4355 - 4364, July, 2015

[34] M. Savaghebi, A. Jalilian, J.C. Vasquez and J. M. Guerrero, "Secondary control scheme for voltage unbalance compensation in an islanded droop-controlled microgrids," *IEEE Transactions on Smart Grid*, vol. 3, no. 99, pp. 1 - 11, 2011

[35] M. Savaghebi, A. Jalilian, J.C. Vasquez and J. M. Guerrero, "Secondary control for voltage quality enhancement in microgrids," *IEEE Transactions on Smart Grid*, vol. 3, no. 4, pp. 1893 - 1902, December, 2012

[36] M. Prodanovic, T. C. Green, "High-quality power generation through distributed control of a power park microgrids," *IEEE Transactions on Industrial Electronics*, vol. 53, pp. 1471 - 1482, October, 2006

[37] J.C. Vasquez, J.M. Guerrero, A. Luna, P. Rodriguez, R. Teodorescu, "Adaptive droop control applied to voltage-source inverters operating in grid-connected and islanded modes," *IEEE Transactions on Industrial Electronics*, vol. 56, no. 10, pp. 4088 - 4096, 2009

[38] Y. W. Li and C.-N. Kao, "An accurate power control strategy for power-electronics-interfaced distributed generation units operating in a low- voltage multibus microgrid," *IEEE Transactions on Power Electronics*, vol. 24, no. 12, pp. 2977 - 2988, 2009

[39] M. Gao, Y. Zhang, C. Jin, M. Chen and Z. Qian, "An improved droop control method for parallel operation of distributed generations in microgrid," *IEEE Applied Power Electronics Conference and Exposition*, pp. 3016 - 3020, 2013

[40] J. Kim, J. M. Guerrero, P. Rodriguez, K. Nam, "Mode adaptive droop control with virtual output impedances for an inverter-based flexible AC microgrid," *IEEE Transactions on Power Electronics*, vol. 26, no. 3, pp. 689 - 701, 2013

[41] J. He, Y. W. Li, "Analysis and design of interfacing inverter output virtual impedance in a low voltage microgrid," *IEEE Energy Conversion Congress and Exposition*, pp. 2857 - 2864, 2010

[42] J. M. Guerrero, N. Berbel, J. Matas, J. L. Sosa, "Droop control method with vitural output impedance for parallel operation of uninterruptible power supply systems in a microgrid," *IEEE Applied Power Electronics Conference*, pp. 1126 - 1132, 2007

[43] J. He, Y. W. Li, "Analysis, design, and implementation of virtual impedance for power electronics interfaced distributed generation," *IEEE Transcations on Industrial Application*, vol. 47, no. 6, pp. 2525 - 2538, 2011

[44] Q. Zhong, G. Weiss, "Synchronverters: Inverters that minic synchronous generators," *IEEE Transactions on Industrial Electronics*, vol. 58, no. 4, pp. 1259 - 1267, 2011

[45] S. M. Ashabani, Y. A. I. Mohamed, "A flexible control strategy for grid-connected and islanded microgrids with enhanced stability using nonlinear microgrid stabilizer," *IEEE Transactions on Smart Grid*, vol. 3, no. 3, pp. 1291 - 1301, 2012

[46] R. Aghatehrani, R. Kavasseri, "Sliding mode control approach for voltage regulation in microgrid with DFIG based wind generations," *IEEE Power and Energy Society General Meeting*, pp. 1 - 8, 2011

[47] S. Dasgupta, S. N. Mohan, S. K. Sahoo and S. K. Panda, "Lyapunov function-based current controller to control active and reactive power flow from a renewable energy source to a generalized three-phase microgrid system," *IEEE Transactions on Industrial Electronics*, vol. 60, no. 2, pp. 799 - 813, 2013

[48] K. T. Tan, P. L. So, Y. C. Chu and M. Z. Q. Chen, "Coordinated control and energy management of distributed generation inverters in a microgrid," *IEEE Transactions on Power Delivery*, vol. 28, no. 2, pp. 704 - 713, 2013

[49] M. Babazadeh, H. Karimi, "A robust two-degree-of-freedom control strategy for an islanded microgrid," *IEEE Transactions on Power Delivery*, vol. 28, no. 3, pp. 1339 - 1347, 2013

[50] A. Kahrobaeian and Y. A. I. Mohamed, "Suppression of interaction fynamics in DG converter-based microgrids via robust system-oriented control approach," *IEEE Transactions on Smart Grid*, vol. 3, no. 4, pp. 1800 - 1811, 2012

[51] A. Kahrobaeian and Y. A. I. Mohamed, "Interactive distributed generation interface for flexible micro-grid operation in smart distribution systems," *IEEE Transactions on Sustainable Energy*, vol. 3, no. 2, pp. 295 - 305, 2012

[52] D. Andrea, R. Dullerud, "Distributed control design for spatially interconnected systems," *IEEE Transactions on Automatic Control*, vol. 23, no. 9, pp. 1478 - 1495, 2003

[53] W. Yu, G. Wen, X. Yu, Z. Wu and J. Lu, "Bridging the gap between complex networks and smart grids," *Journal of Control and Decision*, vol. 1, no. 1, pp. 102 - 114, 2014

[54] A. Bidram, A. Davoudi, F. K. Lewis, J. M. Guerrero, "Distributed cooperative secondary control of microgrids using feedback linearization," *IEEE Transactions on Power System*, vol. 28, no. 3, August, 2013

[55] A. Bidram A. Davoudi, F. K. Lewis, Z. Qu, "Secondary control of microgrids based on distributed cooperative control of multi-agent systems," *IET Generenation, Transmisison & Distribution*, vol. 7, no. 8, pp. 822 - 831, 2013

[56] Q. Shafiee, J. M. Guerrero, J. C. Vasquez, "Distributed secondary control for islanded microgrids- A novel approach," *IEEE Transactions on Power Electronics*, vol. 29, no. 2, pp. 1018 - 1031, February, 2014

[57] John W. Simpson-Porco, Florian Dorfler, Francesco Bullo, "Sychronization and power sharing for droop-controlled inverters in islanded microgrids," *Automatica*, vol. 49, pp. 2603 - 2611, 2013

[58] L.-Y. Lu, C.-C. Chu, "Consensus-based secondary frequency and voltage droop control of virtual synchronous generators for isolated AC microgirds," *IEEE Journal on Emerging and Selected Topics in Circuits and Systems*, vol. 5, no. 3, pp. 443 - 455, 2015

[59] A. Bidram, A. Davoudi, F. L. Lewis, "A multiobjective distributed control framework for islanded AC microgrids," *IEEE Transactions on Industrial Informatics*, vol. 10, no. 3, pp. 1785 - 1798, 2014

[60] B. Singh, K. Al-haddad and A. Chandra, "A review of active filters for power quality improvement," *IEEE Transactions on Industrial Electronics*, vol. 46, no. 5, pp. 960 - 971, 1999

[61] P. T. Cheng, C. Chen, T. L. Lee and S. Y. Kuo, "A cooperative imbalance compensation method for distributed-generation interface converters," *IEEE Transactions on Industrial Application*, vol. 45, no. 8, pp. 2811 - 2820, 2009

[62] M. Hojo, Y. Iwase, T. Funabashi and Y. Ueda, "A method of three phase balancing in microgrid by potovoltaic generation systems," *in Proceedings of Power Electronics and Motion Control Conference*, pp. 2487 - 2491, 2008

[63] M. Savaghebi, A. Jalilian, J.C. Vasquez and J. M. Guerrero, "Autonomous voltage unbalance compensation in an islanded droop-controlled microgrid," *IEEE Transactions on Industrial Electronics*, vol. 60, no. 4, pp. 1390 - 1402, 2013

[64] M. Savaghebi, A. Jalilian, J.C. Vasquez and J. M. Guerrero, "Secondary control for voltage quality enhancement in microgrids," *IEEE Transactions on Smart Grid*, vol. 3, no. 4, pp. 1893 - 1902 , December, 2012

[65] M. Savaghebi, A. Jalilian, J.C. Vasquez and J. M. Guerrero, "Secondary control scheme for voltage unbalance compensation in an islanded droop-controlled microgrids," *IEEE Transactions on Smart Grid*, vol. 3, no.2, pp. 1 - 11 , 2012

[66] V.C. Gungor, D. Sahin, T. Kocak, S. Ergut, "Smart grid technologies: Communication technologies and standards," *IEEE Transactions on Industrial Informatics*, vol. 7, no. 4, pp. 529 - 539, November, 2011

[67] A. Safdarian, M. Fotuhi-Firuzabad, and M. Lehtonen, "A distributed algorithm for managing residential demand response in smart grids," *IEEE Transactions on Industrial Informatics*, vol. 10, no. 4, pp. 2385 - 2393, November, 2014

[68] Y. Xu, W. Liu, "Novel multiagent based load restoration algorithm for micorgrids," *IEEE Transactions on Smart Grid*, vol. 2, no. 1, pp. 152 - 161, 2011

[69] P. Carvalho, P. Correia and L. Ferreira, "Distributed reactive power generation control for voltage rise mitigation in distributed networks," *IEEE Transactions on Power Systems*, vol. 23, no. 2, pp. 766 - 772, 2008

[70] M. R. Irving and M. J. H. Sterling, "Efficient Newton-Raphason algorithm for load-flow calculation in transmission and distribution networks," *in Proceedings of IEE Generations, Transmission Distribution*, vol.134, no.5, pp.325 - 330, 1987

[71] D. Forner, T. Erseghe, S. Tomasin and P. Tenti, "On efficient use of local sources in smart grids with power quality constraints," *in Proceedings of IEEE International Conference on Smart Grid Communication*, USA, pp. 555 - 560, 2010

[72] F. Lu and Y. Y. Hsu, "Fuzzy dynamic programming approach to reactive power/voltage control in a distribution substation," *IEEE Transactions on Power Systems*, vol. 12, no. 2, pp. 681 - 688, 1997

[73] E. Sortomme and M. A. El-Sharkawi, "Optimal power flow for a system of microgrids with controllable loads and battery storage," *in Proceedings of IEEE Power System Conference and Exposition*, pp. 1 - 5, 2009

[74] Z. Luo, W. Ma, M. C. So and S. Zhang, "Semidefinite relaxation of quadratic optimization problems," *IEEE Signal Processing Magzine*, vol. 27, no. 3, pp. 20 - 34, 2010

[75] S. Bolognani and S. Zampieri, "A distributed control strategy for reactive power compensation in smart microgrids," *IEEE Transactions on Automatic Control*, vol. 58, no. 11, pp. 2818 - 2833, 2013

[76] E. D. Anese, H. Zhu and G. B. Giannakis, "Distributed optimal power flow for smart microgrids," *IEEE Transactions on Smart Grid*, vol. 4, no. 3, pp. 1464 - 1475, 2013

[77] D. E. Olivares, A. Mehrizi-Sani, A. H. Etemadi and etc., "Trends in microgrid control," *IEEE Transactions on Smart Grid*, vol. 5, no. 4, pp. 1905 - 1919, 2014

[78] J. -Y Fan, and L. Zhang, "Real-time economic dispatch with line flow and emission constraints using quadratic programming," *IEEE Transactions on Power Systems*, vol. 13, no. 2, pp. 320 - 325, 1998

[79] C. E. Lin, S. T. Chen, and C. L. Huang, "A direct Newton-Raphson economic dispatch," *IEEE Transactions on Power Systems*, vol. 7, no. 3, pp. 1149 - 1154, 1992

[80] T. Guo, M. Henwood, and M. van Ooijen, "An algorithm for combined heat and power economic dispatch," *IEEE Transactions on Power Systems*, vol. 11, no. 4, pp. 1778 - 1784, 1996

[81] H. Liang, B. J. Choi, A. Abdrabou, W. Zhuang, X. Shen, "Decentralized economic dispatch in microgrids via heterogeneous wireless networks," *IEEE Journal on Selected Areas in Communication*, vol. 30, no. 6, pp. 1061 - 1074, 2012

[82] G. Chen, F. L. Lewis, E.-N. Feng, Y.-D. Song, "Distributed optimal active power control of multiple generation systems," *IEEE Transactions on Industrial Electronics*, vol. 62, no. 11, pp. 7079 - 7090, 2015

[83] F. Guo, C. Wen, G. Li and J. Chen, "Distributed economic dispatch for a multi-area power system," *2015 IEEE 10th Conference on Industrial Electronics and Applications*, pp. 620 - 625, Auckland, Newzealand, 2015

[84] Z. Fan, "A distributed demand response algorithm and its application to PHEV charging in smart grids," *IEEE Transactions on Smart Grid*, vol. 3, no. 3, pp. 1280 - 1290, 2012

[85] Z. Zhang and M. Chow, "Convergence analysis of the incremental cost consensus algorithm under different communication network topologies in a smart grid," *IEEE Transactions on Power Systems*, vol. 27, no. 4, pp. 1761 - 1768, November, 2012

[86] S. Yang, S. Tan and J. Xu, "Consensus based approach for economic dispatch problem in a smart grid," *IEEE Transactions on Power Systems*, vol. 28, no. 4, pp. 4416 - 4426, 2013

[87] G. Binetti, A. Davoudi, F. K. Lewis, D. Naso and B. Turchiano, "Distributed consensus-based economic dispatch with transmission losses, " *IEEE Transactions on Power Systems*, vol. 29, no. 4, pp. 1711 - 1720, 2014

[88] G. Binetti, A. Davoudi, D. Naso, B. Turchiano, F. L. Lewis, "A distributed auction-based algorithm for the nonconvex economic dispatch problem," *IEEE Transactions on Industrial Informatics*, vol. 10, no. 2, pp. 1124 - 1132, May 2014

[89] W. Zhang, W. Liu, X. Wang, L. Liu and F. Ferrese, "Online optimal generation control based on constrained distributed gradient algorithm," *IEEE Transactions on Power Systems*, vol. 30, no. 1, pp. 35 - 45, 2015

[90] C. Li, X. Yu and W. Yu, "Optimal economic dispatch by fast distributed gradient," *The 13th Internaional Conference on Control, Automation, Robotics and Vision*, pp. 571 - 576, Singapore, 2014

[91] C. Li, X. Yu and W. Yu, "Distributed Event-triggered scheme for economic dispatch in smart grid," *IEEE Transactions on Industrial Informatics*, vol. 12, no. 5, pp. 1775 - 1785, 2016

[92] M. Alizadeh, X. Li, Z. Wang, A. Scaglione, and R. Melton, "Demandside management in the smart grid: Information processing for the power switch," *IEEE Signal Processing Magzine*, vol. 29, no. 5, pp. 55 - 67, September, 2012

[93] Y. Mou, H. Xing, Z. Lin, M. Fu, "Decentralized optimal demand-side management for PHEV charging in a smart grid," *IEEE Transactions on Smart Grid*, vol. 6, no. 2, pp. 726 - 736, 2015

[94] R. Deng, Z. Yang, M.-Y. Chow, and J. Chen, "A survey on demand response in smart grids: mathematical models and approaches," *IEEE Transactions on Industrial Informatics*, vol. 11, no. 3, pp. 570 - 582, June, 2015

[95] H. Huang, F. Li, and Y. Mishra, "Modeling dynamic demand response using monte carlo simulation and interval mathematics for boundary estimation," *IEEE Transactions on Smart Grid*, vol. 6, no. 6, pp. 2704 - 2713, 2015

[96] M. Alizadeh, X. Li, Z. Wang, A. Scaglione, and R. Melton, "Demand-side management in the smart grid: information processing for the power switch," *IEEE Signal Processing Magzine*, vol. 29, no. 5, pp. 55 - 67, September, 2012

[97] J. Chen, F. N. Lee, A. M. Breipohl, and R. Adapa, "Scheduling direct load control to minimize system operation cost," *IEEE Transactions on Power Systems*, vol. 10, no. 4, pp. 1994 - 2001, 1995

[98] Z. Chen, L. Wu, and Y. Fu, "Real-time price-based demand response management for residential appliances via stochastic optimization and robust optimization," *IEEE Transactions on Smart Grid*, vol. 3, no. 4, pp. 1822 - 1830, 2012

[99] M. Crew, C. Fernando, abd P. Kleindorfer, "The theory of peak-load pricing: A survey," *Journal of Regulatory Economics*, vol. 8, no. 3, pp. 215 - 248, 1995

[100] P. Samadi, A. Mohsenian-Rad, R. Schober, V. W. S. Wong, and J. Jatskevich, "Optimal real-time pricing algorithm based on utility maximization for smart grid," *2010 IEEE International Conference on Smart Grid Communications*, pp. 415 - 420, Gaithersburg, MD, 2010

[101] K. Ma, G. Hu, and C. J. Spanos, "Distributed energy consumption control via real-time pricing feedback in smart grid," *IEEE Transactions on Control System Technology*, vol. 22, no. 5, pp. 1907 - 1914, 2014

[102] Y. Ma, W. Zhang, W. Liu, and Q. Yang, "Fully distributed social welfare optimization with line flow constraint consideration," *IEEE Transactions on Industrial Informatics*, vol. 11, no. 6, pp. 1532 - 1541, 2015

[103] A.-H. M. Rad, Vicent W. S. Wong, J. Jatskevich, R. Schober, and A. Leon-Garcia, "Autonomous demand-side management based on game-theoretic energy consumption scheduling for the future smart grid," *IEEE Transactions on Smart Grid*, vol. 1, no. 3, pp. 320 - 331, 2010

[104] S. Maharjan, Q. Zhu, Y. Zhang, S. Gjessing and T. Basar, "Dependable demand response management in the smart grid: A stackelberg game approach," *IEEE Transactions on Smart Grid*, vol. 4, no. 1, pp. 120 - 132, 2013

[105] M. Ye, G. Hu, and C. J. Spanos, "Distributed extremum seeking for constrained networked optimization and its application to energy consumption control in smart grid," *IEEE Transactions on Control System Technology*, vol. 24, no. 6, pp. 2048 - 2058, November, 2016

[106] C. Godsil and G. Royle, "Algebraic graph theory," *Springer*, 2001

[107] R. Olfati-Saber, J. A. Fax and R. M. Murrary, "Consensus and cooperation in networked multi-agent systems," *Proceeding of the IEEE*, vol.95, no.1, pp. 215 - 233, 2007

[108] M. Zhu, S. Martinez, "Discrete-time dynamic average consensus," *Automatica*, vol. 46, no. 2, pp. 322 - 329, 2010

[109] G. S. Seyboth, D. V. Dimarogonas, and K. H. Johnsson "Event-based broadcasting for multi-agent average consensus," *Automatica*, vol. 49, no. 1, pp. 245 - 252, 2015

[110] A. Y. Kibangou, "Graph Laplacian based matrix design for finite-time distributed average consensus," *in Proceedings of American Control Conference*, pp. 1901 - 1906, 2012

[111] U. Vishkin, "An efficient distributed orientation algorithm (Corresp.)," *IEEE Transactions on Information Theory*, vol. 29, no. 4, pp. 624 - 629, 1983.

[112] R. Aragues, G. Shi, D. V. Dimarogonas, C. Sagues, and K. H. Johansson, "Distributed algebraic connectivity estimation for adaptive event-triggered consensus," *in Proceedings of American Control Conference*, pp. 32 - 37, 2012

[113] Y. Wang, D. Cheng, Y. Hong, and H. Qin, "Finite-time stabilizing excitation control of a synchronous generator," *International Journal of System Science*, vol. 33, no. 1, pp. 13 - 22, 2002

[114] S. Bhat, and D. Bernstein, "Continuous finite-time stabilization of the translational and rotational double integrators," *IEEE Transactions on Automatic Control*, vol. 43, no. 11, pp. 678 - 682, Febraury, 1998

[115] M. Rosier, "Homogeneous Lyapunov function for homogeneous continuous vector filed" *System and Control Letter*, vol. 19, pp. 467 - 473, Febraury, 1992

[116] Y. Hong, J. Huang, and Y. Xu, "On an output feedback finite-time stabilization problem," *IEEE Transactions on Automatic Control*, vol. 46, no. 2, pp. 305 - 309, Febraury, 2001

[117] Albert Y. S. Lam, B. Zhang, D. Tse, "Distributed algorithms for optimal power flow problem," *IEEE Annual Conference on Decision and Control*, Maui, HI, USA, pp. 430 - 437, 2012

[118] Y. Xue, B. Li, and K. Nahrstedt, "Optimal reource allocation in wireless Ad Hoc Networks: A price-based approach," *IEEE Transactions on Mobile Computing*, vol. 5, no. 4, pp. 347 - 364, 2006

[119] Z. Wang, L. Jiang, and C. He, "Optimal price-based power control algorithm in congnitive radio networks," *IEEE Transactions on Wireless Communications*, vol. 13, no. 11, pp. 5909 - 5920, 2014

[120] P. Wan, and M. D. Lemmon, "Distributed network utility maximization using event-triggered augmented Lagrangian methods," *2009 American Control Conference*, St. Louis, MO, USA, pp. 3298 - 3303, 2009

[121] B. Gharesifard, J. Cortes, "Distributed continous-time convex optimization on weight-balanced digraphs," *in Proceedings of IEEE Conference on Decision and Control*, pp. 7451 - 7456, 2012

[122] G. Droge, H. Kawashima, and M. B. Egerstedt, "Continuous-time proportional-integral distributed optimisation for networked systems," *Joural of Control and Decision*, vol. 1, no. 3, pp. 191 - 213, 2014

[123] B. Gharesifard, J. Cortes, "Continuous-time distributed convex optimization on weight-balanced digraphs," *Journal of Control and Decision*, vol. 1 no. 2, pp. 166 - 179, 2014

[124] S. S. Kia, J. Cortes, S. Martinez, "Distributed convex optimization via continuous-time coordination algorithms with discrete-time communication," *Automatica*, vol. 55, pp. 254 - 264, 2015

[125] P. Yi, Y. Hong, and F. Liu, "Distributed gradient algorithm for constrained optimization with application to load sharing in power systems," *System and Control Letter*, vol. 83, pp. 45 - 52, 2015

[126] M. Zhu, S. Martinez, "On distributed convex optimization under inequality and equality constraints," *IEEE Transactions on Automatic Control*, vol. 57, no. 1, pp. 151 - 164, 2012

[127] JC Duchi, A. Agarwal, and M.J. Wainwright, "Dual averaging for distributed optimization: convergence analysis and network scaling," *IEEE Transactions on Automatic Control*, vol. 57, no. 3, pp. 592 - 606, 2012

[128] P. Bianchi, J. Jakubowicz, "Convergence of a multi-agent projected stochastic gradient algorithm for non-convex optimization," *IEEE Transactions on Automatic Control*, vol. 58, no. 2, pp. 391 - 405, 2013

[129] P. Lin, W. Ren and Y. Song, "Distributed multi-agent optimization subject to nonidentical constraints and communication delays," *Automatica*, vol. 65, pp. 120 - 131, 2016

[130] A. Nedic, A. Ozdaglar, P. A. Parrilo, "Constrained consensus and optimization in multi-agent networks," *IEEE Transactions on Automatic Control*, vol. 55, no. 4, pp. 922 - 938, 2010

[131] A. Nedic, A. Ozdaglar, "Distributed subgradient methods for multi-agent optimization," *IEEE Transactions on Automatic Control*, vol. 54, no. 1, pp. 48 - 61, 2009

[132] A. Nedic and D. P. Bertsekas, "Incremental subgradient methods for nondifferentiable optimization," *SIAM Journal on Optimization*, vol. 12, no. 1, pp. 109 - 138, 2001

[133] B. Johansson, M. Rabi and M. Johansson, "A randomized incremental subgradient method for distributed optimization in networked systems," *SIAM Journal on Optimization*, vol. 20, no. 3, pp. 1157 - 1170, 2009

[134] S. Boyd, N. Parikh, E. CHu, B. Peleato, and J. Eckstein, "Distributed optimization and statistical learning via the alterbating direction method of multipliers," *Foundations and Trends in machine Learing*, vol. 3, no. 1, pp. 1 - 122, 2010

[135] H. Wang, Y. Gao, Y. Shi, and R. Wang, "Group-based alternating direction method of multipliers for distributed linear classification," *IEEE Transactions on Cybernetics*, published on line, DOI: 10.1109/TCYB.2016.2570808

[136] Y. Ouyang, Y. Chen, G. Lan and E. Pasiliao Jr., "An accelerated linearized alterbating direction method of multipliers," *SIAM Journal on Imaging Sciences*, vol. 8, no. 1, pp. 644 - 681, 2015

[137] N. Pogaku, M. Prodanovic, T. C. Green, "Modeling, analysis and testing of autonomous operation of an inverter-based microgrid," *IEEE Transactions on Power Electronics*, vol. 22, no. 2, pp. 613 - 625, March 2003

[138] K. Hatipoglu, I. Fidan, G. Radman, "Investigating effect of voltage changes on static ZIP load model in a microgrid enviroment," *In North American Power Symposium (NAPS)*, 2012, pp. 1 - 5, September, 2012

[139] A.R. Bergen and V. Vittal, "Power system analysis," *Prentice Hall*

[140] Johannes Schiffer, Ortega Romeo, Alessandro Astolfi, Jorg Raisch, Tevfik Sezi, "Conditions for stability of droop-controlled inverter-based microgrids," *Automatica*, vol. 50, no. 10, pp. 2457 - 2469, October, 2014

[141] Zhao Wang, Meng Xia and M. Lemmon, "Voltage stability of weak power distribution networks with inverter connected sources," *2013 American Control Conference*, Washington DC, USA, pp.6577 - 6582, 2013

[142] H. Bouattour, "Distributed secondary control in microgrids," *Universitat Stuttgart*, 2013

[143] D. Cheng, T. J. Tarn, and A. Isidori, "Global external linearization of nonlinear systems via feedback," *IEEE Transactions on Automatic Control*, vol. 30, no. 8, pp. 808 - 811, 1985

[144] J. Slotine, W. Li, "Applied nonlinear control," *Upper Saddle River*, NJ, USA: Prentice-Hall, 2009

[145] J. Hu, Y. Hong, "Leader-following coordination of multi-agent systems with coupling time delays," *Physica A*, vol. 374, no. 2, pp. 853 - 863, 2007

[146] M. Andreasson, D. V. Dimarogonas, H. Sandberg, and K. H. Johansson, "Distributed control of networked dynamic systems: Static feedback, integral action and consensus," *IEEE Transactions on Automatic Control*, vol. 59, no. 7, pp.1750 - 1765, July 2014

[147] Y. Chen, J. Lu, and Z. Lin, "Consensus of discrete-time multi-agent systems with transmission nonlinearity," *Automatica*, vol. 49, no. 6, pp. 1768 - 1775, Jun. 2013

[148] C. T. Chen, "Linear System Theory and Design," *Oxford University Press*, 3rd Edition, 1999

[149] A. Siddique, G. S. Yadava, B. Singh "Effects of voltage unbalance on induction motors," *2004 IEEE International Symposum on Electrical Insulation*, pp. 26 - 29, 2004

[150] B. Singh, K. Al-haddad and A. Chandra, "A review of active filters for power quality improvement," *IEEE Transactions on Industrial Electronics*, vol. 46, no. 5, pp. 960 - 971, 1999

[151] M. Park, M. Yim, "Distributed control and communication fault tolerance for the CKBot," *ASME/IFTOMM international conference on reconfigurable mechanisms and robots (REMAR 2009)*, London, UK, pp. 682 - 688, 2009

[152] H. Liang, B. J. Choi, W. Zhuang, and X. Shen, "Stability enhamcement of decentralized inverter control through wireless communications in micorgrids," *IEEE Transactions on Smart Grid*, vol. 4, no. 1, pp. 321 - 331, 2013

[153] Y. Xu, W. Liu, "Novel multiagent based load restoration algorithm for micorgrids," *IEEE Transactions on Smart Grid*, vol. 2, no. 1, pp. 152 - 161, 2011

[154] W. Gu, W. Liu, J. Zhu, B. Zhao, Z. Wu, Z. Luo, and J. Yu, "Adaptive decentralized under-frequency load shedding for islanded smart distribution networks," *IEEE Transactions on Sustainable Energy*, vol. 5, no. 3, pp. 886 - 895, 2014

[155] W. Liu, W. Gu, W. Sheng, X. Meng, Z. Wu, and W. Chen, " Decentralized multi-agent system-based cooperative frequency control for autonomous microgrids with communication constraints," *IEEE Transactions on Sustainable Energy*, vol. 5, no. 2, pp. 446 - 456, 2014

[156] Y. Xu, W. Liu and J. Gong, "Stable multi-agent-based load shedding algorithm for power systems," *IEEE Transactions on Power Systems*, vol. 26, no. 4, pp. 2006 - 2014, 2011

[157] "Electric power systems and equipment-voltage ratings (60Hertz)," *ANSI Stand. Publ. no. ANSI C84.1-1995*

[158] K-J Lee, J-P Lee, D. Shin, D-W Yoo, "A novel grid synchronization PLL method based on adaptive low-pass notch filter for grid-connected PCS," *IEEE Transactions on Industrial Electronics*, vol. 61, no. 1, pp. 292 - 301, 2014

[159] J. Chen, J. Chen and C. Gong, "On optimizing the aerodynamic load acting on the turbine shaft of PMSG-based direct-drive wind energy conversion system, "*IEEE Transactions on Industrial Electronics*, vol. 61, no. 8, pp. 4022 - 4031, 2014

[160] J. Chen, J. Chen and C. Gong, "New overall power control strategy for variable-speed fixed pitch wind turbines within the whole wind velocity range, "*IEEE Transactions on Industrial Electronis*, vol. 60, no. 7, pp. 2652 - 2660, 2013

[161] Xian Liu, Wilsun Xu, "Economic load dispatch constrained by wind power availability: A here-and-now approach," *IEEE Transactions on Sustainable Energy*, vol. 1, no. 1, pp. 2 - 9, 2010

[162] X. Liu, "Economic load dispatch constrained by wind power availability: A wait-and-see approach," *IEEE Transactions on Smart Grid*, vol. 1, no. 3, pp. 347 - 355, December, 2010

[163] J. Hetzer, D. C. Yu and K. Bhattarai, "An economic dispatch model incorporating wind power," *IEEE Transactions on Energy Conversion*, vol. 23, no. 2, pp. 603 - 611, 2008

[164] X. Liu, W. Xu, "Minimum emission dispatch constrained by stochastic wind power availability and cost," *IEEE Transactions on Power Systems*, vol. 25, no. 3, pp. 1705 - 1713, 2010

[165] Y. Zhang, N. Gatsis, and G. B. Giannakis, "Robust energy management for microgrids with high-penetration renewables," *IEEE Transactions on Sustainable Energy* , vol. 4, no. 4, pp. 944 - 953, 2013

[166] J. P. Barton, D. G. Infield, "Energy storage and its use with intermittent renewable energy," *IEEE Transactions on Energy Conversion*, vol. 19, no. 2, pp. 441 - 448, 2004

[167] Allen J. Wood, Bruce F. Wollenberg, and Gerald B. Sheble, "Power Generation, Operation and Control, 3rd Edition," *John Wiley & Sons, Inc.*, 2012

[168] G. Li, C. Wen, W. Zheng, and Y. Chen, "Identification of a class of nonlinear autoregressive models with exogenous inputs based on kernel machines," *IEEE Transactions on Signal Proceesing*, vol. 59, no. 5, pp. 2146 - 2159, 2011

[169] D. C. Walters, G. B. Sheble, "Genetic algorithm solution of economic dispatch with valve point loading," *IEEE Transactions on Power Systems*, vol. 8, no. 3, pp. 1325 - 1332, 1993

[170] I. G. Damousis, A. G. Bakirtzis, and P. S. Dokopoulos, "Network-constrained economic dispatch using real-coded genetic algorithm," *IEEE Transactions on Power Systems*, vol. 18, no. 1, pp. 198 - 205, 2003

[171] Z -L. Gaing, "Particle swarm optimization to solving the economic dispatch considering the generator constraints," *IEEE Transactions on Power Systems*, vol. 18, no. 3, pp. 1187 - 1195, 2003

[172] A. Mohammadi, M. Varahram, and I. Kheirizad, "Online solving of economic dispatch problem using neural network approach and comparing it with classical method," *in Proc. Int. Conf. Emerging Technologies (ICET06)*, pp. 581 - 586, 2006

[173] S. Boyd, and L. Vandenberghe, "Convex Optimization," *Cambridge Univeristy Press*, 2004

[174] S. Boyd, P. Diaconis, and L. Xiao, "Fastest mixing Markov chain on a graph," *SIAM Review*, vol. 46, pp. 667 - 689, 2004

[175] P. Giselsson, M. D. Doanm T. Keviczky, B. D. Schutter, A. Rantzer, "Accelerated gradient methods and dual decomposition in distributed model predictive control," *Automatica*, vol. 49, no. 3, pp. 829 - 833, 2013

[176] E. Ghadimi, I. Shames, M. Johansson, "Multi-step gradient methods for networked optimization," *IEEE Transactions on Signal Processing*, vol. 61, no. 21, pp. 5417 - 5429, 2013

[177] P. A. Lipka, R. P. O'Neill, S. Oren, "Developing line current magnitude constraints for IEEE test problems - Optimal power flow paper 7 ," 2013

[178] P. Palensky, D. Dietrich, "Demand side management: Demand response, intelligent energy systems, and smart loads," *IEEE Transactions on Industrial Informatics*, vol. 7, no. 3, pp. 381 - 388, 2011

[179] Y. Mou, H. Xing, Z. Lin, M. Fu, "Decentralized optimal demand-side management for PHEV charging in a smart grid," *IEEE Transactions on Smart Grid*, vol. 6, no. 2, pp. 726 - 736, 2015

[180] R. Deng, Z. Yang, M.-Y. Chow, and J. Chen, "A survey on demand response in smart grids: Mathematical models and approaches," *IEEE Transactions on Industrial Informatics*, vol. 11, no. 3, pp. 570 - 582, June, 2015

[181] H. Huang, F. Li, and Y. Mishra, "Modeling dynamic demand response using monte carlo simulation and interval mathematics for boundary estimation," *IEEE Transactions on Smart Grid*, vol. 6, no. 6, pp. 2704 - 2713, 2015

[182] M. Alizadeh, X. Li, Z. Wang, A. Scaglione, and R. Melton, "Demandside management in the smart grid: Information processing for the power switch," *IEEE Signal Processing Magzine*, vol. 29, no. 5, pp. 55 - 67, September, 2012

[183] P. Tarasak, "Residential energy consumption scheduling: A coupled-constraint game approach," *IEEE Transactions on Smart Grid*, vol. 5, no. 3, pp. 1340 - 1350, May, 2014

[184] K. Ma, G. Hu, and C. J. Spanos, "A cooperative demand response scheme using punishment mechanism and application to industrial refrigerated warehouses," *IEEE Transactions on Industrial Informatics*, vol. 11, no. 6, pp. 1520 - 1531, 2015

[185] B. Chai, J. Chen, Z. Yang, and Y. Zhang, "Demand response management with multiple utility companies: A two level game approach," *IEEE Transactions on Smart Grid*, vol. 5, no. 2, pp. 722 - 731, 2014

[186] R. Deng, Z. Yang, J. Chen, N. R. Asr, and M.-Y. Chow, "Residential energy consumption scheduling: A coupled-constraint game approach," *IEEE Transactions on Smart Grid*, vol. 5, no. 3, pp. 1340 - 1350, May, 2014

[187] W. Zhang, Y. Xu, W. Liu, C. Zhang, "Distributed online optimal energy management for smart grids," *IEEE Transactions on Industrial Informatics*, vol. 11, no. 3, pp. 717 - 727, June, 2015

[188] F. Guo, C. Wen, J. Mao, J. Chen, and Y.-D. Song, "Distributed cooperative secondary control for voltage unbalance compensation in an islanded microgrid," *IEEE Transactions on Industrial Informatics*, vol. 11, no. 5, pp. 1078 - 1088, October, 2015

[189] M. Wang, D. P. Bertsekas, "Incremental constraint projection-proximal methods for nonsmooth convex optimization," *Report LIDS-P-2907*, Laboratory for Information and Decision Systems, Massachusetts Institute of Technology, Cambridge, MA, 2013

[190] K. Herter, P. McAuliffe, and A. Rosenfeld, "An exploratory analysis of California residential customer response to critical peak pricing of electricity," *Energy*, vol. 32, no. 1, pp. 25 - 34, January, 2007

[191] J. Manyika, M. Chui, B. Brown, J. Bughin, R. Dobbs, C. Roxburgh, A. H. Byers, "Big data: The next frontier for innovation, competition, and productivity," *McKinsey Global Institute*, 2011

[192] H. Jia, N. Guangyu, S. T. Lee and P. Zhang, "Study on the impact of time delay to power system small signal stability," *in Proceedings of IEEE Mediterranean Electrotechnical Conference*, pp. 1011 - 1014, 2006

[193] S. Qiang, A. Haiyun, W. Wei and M. Zhihui *ect.*, "An improved power system stability criterion with multiple time delays," *in Proceedings of IEEE PES Gernal Meeting*, pp. 1 - 7, 2009

[194] F. Milano, M. Anghel, "Impact of time delays on power system stability," *IEEE Transactions on Circuits and Systems*, vol. 59, no. 4, pp. 889 - 900, 2012

[195] P. Wan and M. D. Lemmon, "Optimal power flow in microgrids using event-triggered optimization," *in Proceedings of American Control Conference*, pp. 2521 - 2526, 2010

[196] Y. Fan, G. Hu, M. Egerstedt, "Distributed reactive power sharing control for microgrids with event-triggered communication," *IEEE Transactions on Control System Technology*, vol. 25, no. 1, pp. 118 - 128, 2017

Index